Power Systems

Electrical power has been the technological foundation of industrial societies for many years. Although the systems designed to provide and apply electrical energy have reached a high degree of maturity, unforeseen problems are constantly encountered, necessitating the design of more efficient and reliable systems based on novel technologies. The book series Power Systems is aimed at providing detailed, accurate and sound technical information about these new developments in electrical power engineering. It includes topics on power generation, storage and transmission as well as electrical machines. The monographs and advanced textbooks in this series address researchers, lecturers, industrial engineers and senior students in electrical engineering.

Power Systems is indexed in Scopus

More information about this series at https://link.springer.com/bookseries/4622

Lei Ding · Qing-Long Han · Boda Ning

Distributed Control and Optimization of Networked Microgrids

A Multi-Agent System Based Approach

Springer

Lei Ding
Institute of Advanced Technology
Nanjing University of Posts
and Telecommunications
Nanjing, China

Qing-Long Han
School of Science, Computing
and Engineering Technologies
Swinburne University of Technology
Melbourne, VIC, Australia

Boda Ning
Swinburne University of Technology
Melbourne, VIC, Australia

National Natural Science Foundation of China
62073171
Natural Science Foundation of Jiangsu
BK20200744
Australian Research Council Discovery Project
DP160103567

ISSN 1612-1287 ISSN 1860-4676 (electronic)
Power Systems
ISBN 978-3-030-95028-6 ISBN 978-3-030-95029-3 (eBook)
https://doi.org/10.1007/978-3-030-95029-3

This Springer imprint is published by the registered company Springer Nature Switzerland AG
The registered company address is: Gewerbestrasse 11, 6330 Cham, Switzerland

Preface

Distributed renewable energy resources, such as wind turbines and photovoltaic panels, have been well exploited and widely penetrated into traditional power grids due to environmental and economic concerns. This leads the traditional power grids to experiencing a huge transformation towards smart grids, especially in terms of power control and management. In order to locally absorb the generation of distributed and renewable energy resources and fully meet the high-reliability demand for various types of loads, a microgrid has been put forward as a promising technique for modern power grids. It is commonly recognized that a microgrid is a small power network or system which is composed of various units, such as distributed generations, loads and energy storages. In general, a networked microgrid can be viewed as a multi-agent system, where each microgrid can be viewed as an agent. The agents can be connected physically by transmission lines, while they can communicate with each other via communication networks.

The large-scale penetration of distributed and renewable energy generation and various types of demands inevitably introduces multiple hybrid uncertainties in both physical and cyber layers, seriously affecting the stability of power systems. On the other hand, with the increasing number of distributed and renewable energy generation and new-style loads in the microgrid, it becomes costly or even impractical to utilize a centralized control and optimization strategy by collecting all the information of units to implement the management of power. By contrast, it is more desirable for microgrids to adopt a distributed control and optimization scheme due to its advantages in improving reliability, efficiency, security and economy of systems. Note that distributed schemes require necessary support from communication networks, implying that various communication constraints, such as communication delays, limited bandwidth and cyberattacks, should be fully taken into account when designing distributed control and optimization schemes for microgrids. Therefore, it is an attractive yet challenging issue for microgrids on how to develop suitable distributed control and optimization schemes which can accommodate the effects of communication constraints.

This monograph aims to provide some techniques and methods for control and optimization of networked microgrids in a multi-agent system-based framework. By establishing a direct connection between multi-agent systems and networked microgrids, some distributed coordinated control and optimization schemes with consideration of communication constraints are presented to ensure reliable, efficient, secure and economic operation of networked microgrids. The monograph will contribute to stimulating extensive interest of researchers in electrical and control fields, and potentially helping electrical engineers with conducting the advanced techniques for improving power system reliability, robustness and security.

Structure and readership. This monograph is concerned with multi-agent system-based distributed control and optimization of networked microgrids. In Chap. 1, a basic background of microgrids is first presented and a brief introduction of recent advances in distributed control and optimization of networked microgrids is given. In Chap. 2, a network-based coordinated control framework is proposed, where network-induced delays are taken into account. Under this framework, a distributed leader-following consensus control scheme is designed to accommodate the effects of network-induced delays. Moreover, distributed event-triggered consensus issues are investigated in Chap. 3, where the relationship between communication resources and the control performance can be explicitly characterized. In Chaps. 4, 5 and 6, by considering the effects of network uncertainties, such as communication delays, limited communication resources and cyberattacks, multi-agent system-based distributed control schemes are tailored to solve the fundamental issues of networked microgrids, such as distributed frequency regulation, voltage regulation and active power sharing/load sharing. Distributed optimization of networked microgrids/smart grids is dealt with in Chaps. 7 and 8. In Chap. 7, a distributed optimal control algorithm for DC microgrids is presented to achieve a trade-off between voltage regulation and load sharing. In Chap. 8, distributed energy management of smart grids is realized by maximizing the social welfare of generation and demands.

Nanjing, China — Lei Ding
Melbourne, Australia — Qing-Long Han
Melbourne, Australia — Boda Ning

Acknowledgments

We would like to acknowledge (i) Profs. Ge Guo, Le Yi Wang, George Yin, Wei Xing Zheng, Dong Yue, Chun-Xia Dou, Fuwen Yang, Chen Peng, Xian-Ming Zhang, Derui Ding, Dr. Xiaohua Ge and Dr. Eyad Sindi, for collaborations on related work in this monograph; (ii) support from the National Natural Science Foundation of China under Grant 62073171, the Natural Science Foundation of Jiangsu Grant BK20200744, the Jiangsu Specially-Appointed Professor Grant RK043STP19001 and the Australian Research Council Discovery Project under Grant DP160103567; and (iii) Mr. Anthony Doyle, Executive Editor, Engineering, Springer, 236 Gray's Inn Road, Floor 6, London WC1X 8HL, UK, for his encouragement to write this monograph.

Contents

Acronyms

A^{-1}	Inverse matrix of an invertible matrix A
A^T	Transpose matrix of a matrix A
$A \geq 0$	Matrix A is symmetric positive semi-definite
$A > 0$	Matrix A is symmetric positive definite
$A \leq 0$	Matrix A is symmetric negative semi-definite
$A < 0$	Matrix A is symmetric negative definite
$\|a\|$	Absolute value (or modulus) of a scalar a
$\\|x\\|$	2-norm of a real vector x
$\\|x\\|_p$	p-norm of a real vector x
$\mathrm{diag}\{\cdot\}$	A diagonal matrix
$\mathbb{R}$	Set of real numbers
$\mathbb{R}^+$	Set of positive real numbers
$\mathbb{R}^m$	Set of m-dimensional real vectors
$\mathbb{R}^{n\times m}$	Set of $n \times m$ real matrices
sign	Signum function
$0_{n\times m}$	Zero matrix of dimension $n \times m$
$\mathbf{1}_n$	Column vector of all ones with n entries
$\mathbf{0}$	Column vector of all zeros
$\boldsymbol{I}_n$	$n \times n$ identity matrix
lim	Limit
max	Maximum
min	Minimum
inf	Infimum

Chapter 1
Introduction

With the ever-increasing demand for electricity, the exhaustion of traditional energy resources (e.g., coal, fossil oil, gas) and environmental deterioration have been becoming the major concern for the modern society. This stimulates the rapid development of green and renewable energy resources. Undoubtedly, smart grids can provide a necessary and important support to realize the transformation of high-efficient, low-carbon and clean energy. Currently, many countries and regions such as China, the United States, Europe and Canada have invested mounts of human and material resources to develop such a novel smart grid. For example, in the "China Manufacturing 2025" plan, smart grids have been listed as one of key breakthrough strategic domains and become a critical mission for "Internet+Smart Energy", aiming fundamentally to make a significant promotion of energy transformation and accomplish the strategic plan of energy development.

1.1 Microgrids

In the context of energy transformation, a microgrid, as an important basis of smart grids, has been put forward to offer a promising paradigm that can achieve the large-scale penetration of distributed and renewable energy resources and meet the high-reliability demand for various types of loads. In general, a micorgrid is a small power network which consists of distributed generations (DGs) (e.g. diesel generator, gas turbine, photovoltaic panel and wind turbine), energy storages (ESs) and loads. Micorgrids can operate in either a grid-connected or an islanded mode, and they can carry out a seamless transition between these two operating modes. For example, when a fault happens to the main grid, the micorgrid can be disconnected from the main grid and operate in an islanded mode to maintain the energy supply to local demand and ensure the stability of systems.

L. Ding et al., *Distributed Control and Optimization of Networked Microgrids*, Power Systems, https://doi.org/10.1007/978-3-030-95029-3_1

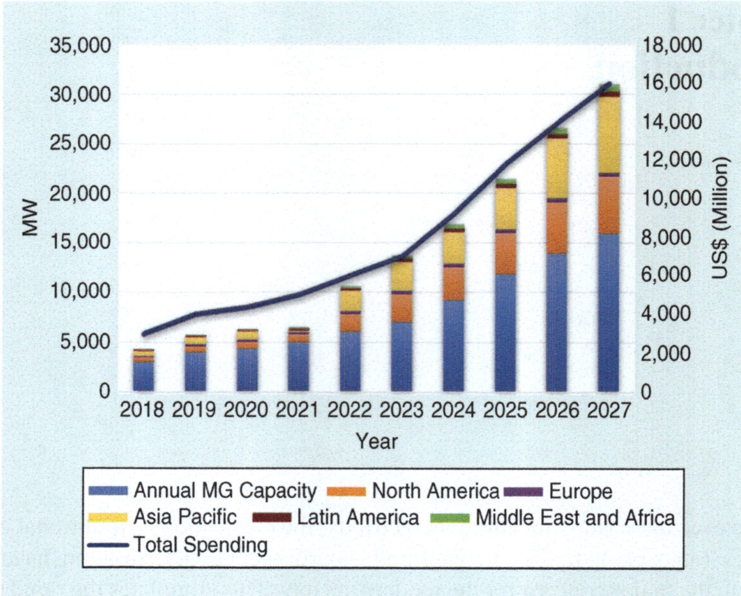

Fig. 1.1 The annual micorgrid capacity and world market predict [1]

Compared with traditional power grids, the key characteristics of micorgrids are to locally achieve the absorption and balance of distributed energy, significantly showcasing the advantages, in terms of: (i) improving the security and reliability of power grids; (ii) promoting the integration of renewable and distributed energy resources into power grids; (iii) enhancing the quality of electricity and reliability of electricity supply; and (iv) reducing the loss of power networks. Therefore, micorgrids have great potentials with wide applications in industrial and commercial regions, urban and rural regions and remote regions. In recent years, investigations for micorgrid techniques have been being undertaken in a growing rate all over the world. According to the report from Navigant Research, in which a market predict for annual micorgrid capacities and consumption in the global world from 2018 to 2027 has been made (see Fig. 1.1), it was shown that the total energy production of micorgrids will reach 33 GW approximately by the end of 2027 [1].

1.2 Hierarchical Control Structures

Due to the integration of renewable and distributed energy resources (e.g., wind, solar) and various new businesses and demands (e.g., charging plies of electric cars), both physical and information technologies in power systems have been experiencing a huge revolution so as to accommodate multiple and hybrid uncertainties that

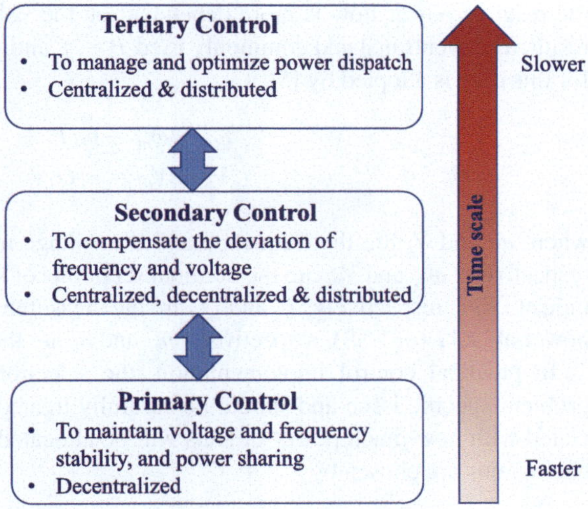

Fig. 1.2 A schematic diagram of hierarchical structure for micorgrids

may be caused by the intermittency of renewable energy generations (restricted by nature conditions) and the rapid changes of demands. Besides, the utilization of advanced communication and computation technologies and the relative openness of information networks make micorgrids vulnerable to the invasion of cyberattacks from adversaries, probably leading to secure risks. As a result, such uncertainties in both physical and cyber layers pose new challenges and difficulties in implementing control and management for micorgrids in reliable, secure, efficient and economic fashions. Operations of control and management mainly concentrate on microgrid stability and optimization involving frequency synchronization, voltage regulation and power dispatch, etc., while maintaining supply and demand power balance over the entire microgrid. In order to address these issues, a standardized and hierarchical control architecture has been put forward to provide an effective and widely adopted solution which enables the multi-objective tasks to be separately achieved in different layers and time scales. As illustrated in Fig. 1.2, the hierarchical control architecture includes primary control, secondary control and tertiary control [2–4], which are explained and reviewed below in detail.

1.2.1 Primary Control

As the first control layer, the primary control is responsible to guarantee stability of frequency and voltage as well as proportional power sharing among multiple DGs. This can be achieved by using a droop control scheme which acts as a virtual impedance loop to provide adjustable references of frequency and voltage to the inner current and voltage control loops. For an AC microrid with dominantly inductive lines, the active power flow relies more on the phase angle of bus, while the reactive power flow is more dependent on the voltage magnitude of bus. As a result, a conventional and commonly used $P - f$ and $Q - V$ droop control scheme for this case is adopted by [5]

$$\omega_i = \omega_{ni} - m_i P_i \tag{1.1a}$$

$$V_i = V_{ni} - n_i Q_i \tag{1.1b}$$

where ω_i and V_i are the frequency and the voltage magnitude of DG i (or ESS), respectively; ω_{ni} and V_{ni} are the nominal set points of frequency and voltage magnitude at DG i, respectively; P_i and Q_i are the measurement of active and the reactive power at DG i (or ESS), respectively; m_i and n_i are the droop coefficients.

In practical control implementation, the selection of droop coefficients is a problem-specific issue and should be carefully treated since they are closely associated with how much power of DGs will be allocated to loads. Usually, the droop coefficients are chosen by

$$\omega_i = \frac{\Delta\omega_i}{P_i^M} = \frac{\omega_i^{max} - \omega_i^{min}}{P_i^M} \tag{1.2a}$$

$$n_i = \frac{\Delta V_i}{Q_i^M} = \frac{V_i^{max} - V_i^{min}}{Q_i^M} \tag{1.2b}$$

where $\Delta\omega_i$ and ΔV_i are the maximum allowable operation limits of frequency and voltage at DG i, with ω_i^{max} and ω_i^{min} being the allowable maximum and minimum frequencies, respectively; P_i^M and Q_i^M are the maximum active and reactive power of DG i, respectively; V_i^{max} and V_i^{min} are the allowable maximum and minimum voltage magnitudes of DG i, respectively. When $\Delta\omega_i$ and ΔV_i in (1.2) are all identical for any DG i, it is easy to know that the loads of the microgrid will be fairly shared among DGs in proportion to their generation capacity while guaranteeing the frequency/ voltage operational limits of DGs.

In addition, different from AC microgrids, DC microgrids do not include any reactive power and frequency control issues. Hence, the following $V - I$ droop control is used for a DC microgrid

$$V_i = V_{ni} - R_i I_i \tag{1.3}$$

where V_i and I_i are the output voltage and current magnitude of DG i, respectively.

It is well known that the droop control (1.1) can realize frequency synchronization, voltage stability and power quality of microgrids. Moreover, note that the droop control is performed in a decentralized way, i.e., using only local measurement without requiring any communication among DGs. Thus, the droop control can be simple and efficient to implement control and management of microgrids. For a lossless microgrid, using the droop controller (1.1) can ensure accurate active power sharing while synchronizing the frequency of each DG. However, it should be noted that there is a deviation of synchronized frequency from the desired value, which implies that the droop control has the drawback of being unable to ensure high precision of frequency regulation and active power sharing. Besides, due to the effects of line impedances, it is not possible to achieve the accurate reactive power sharing among DGs which are even all identical and connected in parallel.

In addition to the P/Q droop control in (1.1), some different droop control methods have been employed for different kinds of microgrids. For example, when a microgrid uses a dominantly resistive transmission line, an alternative $V - P$ and $f - Q$ droop control scheme was introduced in [6], where the active and reactive power can be decoupled by the virtual impedance as well. Considering that the linear-like droop controller (1.1) does not explicitly reflect the quadratically nonlinear nature which is inherent in reactive power flows, a slightly modified version of the voltage droop control (1.1), termed as the quadratic voltage droop control, was proposed in [7]. Detailed reviews on various droop control techniques can be found in the recent surveys [2, 3].

1.2.2 Secondary Control

In order to compensate the deviations of frequency and voltage produced by the primary droop control, a secondary control level is necessarily introduced. Traditionally, the secondary control is performed in a centralized way by collecting all the information from DGs in the microgrid through a communication system. Then, based on the information, the secondary controller will generate a compensating control signal and send it to the primary control of each DG in the microgrid. However, the centralized secondary control requires one-to-all communication, thus causing excessive usage of communication resources. Furthermore, a single point of failure in the communication system will probably lead to instability of microgrids. In order to overcome these shortcomings, a control scheme was provided in [8] for secondary frequency and voltage regulation as well as active and reactive power sharing. Unfortunately, the method used in [8] still requires high communication which enables each DGs to exchange their information with all other DGs in the microgrid. As a result, to reduce utilization of communication resources and improve reliability of microgrids, some distributed secondary controllers for frequency regulation/active power sharing and voltage regulation/reactive power sharing have been proposed for AC microgrids and DC micorgrids, which are briefly reviewed as follows.

(1) Distributed secondary control for AC microgrids. To achieve frequency regulation and active power sharing simultaneously, distributed-averaging proportional-integral (DAPI) frequency and voltage control algorithms were presented in [9]. It is shown that, the distributed secondary frequency controller proposed in [9] only needs to exchange the data measurement among neighbors via a communication network, thus showing its remarkable advantages, in terms of less consumption of communication and better robustness to communication link failures. A distributed secondary voltage control scheme was presented in [10], where the voltage regulation problem was transformed into the synchronization problem by using a feedback lineariza-

tion technique. In [11], a consensus-based distributed voltage control scheme was employed to ensure accurate reactive power sharing and voltage stability. In [12], an optimal distributed secondary control approach was provided for frequency/active power and voltage/reactive power control in an islanded microgrid. Regarding uncertainty of communication links, a robust distributed secondary control scheme was proposed in [13] to achieve voltage/frequency regulation and active power sharing. To guarantee fast convergence rates, a distributed finite-time frequency and active power sharing control strategy was presented in [14], where the control inputs were bounded to avoid transient overshoot issues.

(2) Distributed secondary control of battery energy storage systems (BESSs) for AC microgrids. Some efforts have been made on distributed secondary control for multiple BESSs in AC microgrids [15–20]. In [15], a distributed consensus-based optimal control algorithm was introduced to achieve dynamic energy level balancing between BESSs. Based on a simplified dynamics of BESSs, a distributed control scheme was proposed to synchronize energy levels of BESSs [16]. It is shown that the controller gains in [15, 16] have a lower bound depending on the smallest nonzero eigenvalue of the Laplacian matrix of a communication topology. Different from [15, 16], a fully distributed consensus control strategy without requiring any information of a communication topology was designed for droop-based microgrids with BESSs [17], where independent control inputs for energy levels and power sharing were used. Note that all the aforementioned results only focus on the case that the dynamics of all the BESSs are identical, but they cannot apply to the heterogeneous BESSs. Based on this observation, the control method in [17] was extended [18] to deal with the case of heterogeneous BESSs. Considering the heterogeneous nature of BESSs, a multi-agent cooperative control strategy was proposed to ensure the SoC balancing between heterogeneous BESSs [19]. It is worthwhile to mention that the distributed control schemes proposed in [15–19] just ensure asymptotic stability convergence. In practice, it is more desirable and significant to develop a distributed finite-time control scheme due to its fast convergence, better performance and robustness [21]. In [20], a distributed finite-time control scheme for heterogeneous BESSs in a droop-controlled microgrid was presented.

(3) Distributed secondary control in DC microgrids. The modelling of consensus-based distributed hierarchical control was studied [22]. In [23], a distributed control strategy was presented to ensure proportional load sharing while improving voltage regulation in low voltage DC microgrids. A distributed primary/secondary control framework for DC microgrids was established to achieve load sharing and voltage regulation [24].

1.2.3 Tertiary Control

As the highest level of the hierarchical control framework for networked microgrids, the tertiary control is responsible for economically executing operation of micrigrids and efficiently managing the power flow between the microgrid and the main grid.

This level is often operated in a slowest time (minutes or hours) and schedules the optimal plan of power generation based on the predictions of loads grid tariff and environment in the future (perhaps in the next hours and day). Traditionally, the tertiary control level involves two parts: economic dispatch (ED) and demand response (DR). ED focuses on a generation side by minimizing the total generation cost subject to physical constraints of each generator. In contrast, DR aims to maximize the overall benefit of consumers on a demand side. In most of the existing research work, these two operations are individually investigated, e.g., see [5, 25–27] for ED and [28–32] for DR. In practical implementation, however, these control decisions interact and become intertwined. For example, the power allocation of distributed generators executed by ED has an impact on the electricity market price. This in turn affects the operation of DR [33]. Therefore, it is beneficial to investigate integrated energy management by considering both ED and DR in a unified framework, which refers to a social welfare maximization (SWM) problem.

Early results on weighted and constrained consensus methods were reported in [34, 35] for achieving the balanced power generation. Consensus-based algorithms for ED problems were developed [25–27]. A distributed bisection algorithm based on consensus like an iterative method was proposed to solve an ED problem [36]. Conditioned on the fast convergence to optimality, a finite time ED algorithm was given [37], where the initial power balance was needed. For demand side management, a distributed DR algorithm was presented [30] and applied to plug-in hybrid electric vehicle (PHEV) charging in smart grids. In [32], a distributed DR problem was solved by using a price-based scheme with its price still determined by a central utility that collects information on all demands. It is worth mentioning that all the aforementioned results only considered an unilateral case for generation or demand. To efficiently accomplish energy management for both generation and demand, an incremental welfare distributed consensus-based algorithm was presented [38]. The scheme was further extended in [39] to include transmission loss and directed communication topologies. By using an optimal control law, an SWM problem was addressed in [40], where generator and load adjustment rates can be minimized.

1.3 Communication Constraints

Distributed control strategies require necessary supports from communication networks to implement information exchange among neighbors. While the utilization of communication networks can bring some remarkable benefits in improving the reliability, efficiency and scalability of systems, it inevitably introduces a variety of communication constraints, especially due to the limitation of communication resources. These communication constraints may result in detrimental consequences such as long latency, increased packet loss and reduced throughput, thereby inevitably degrading the system stability, performance and reliability. Therefore, it is very essential to investigate the impacts of communication constraints on distributed control and optimization strategies for micorgrids. In what follows, some existing works, which take into account the effects of communication delays, event-triggered communication and cyberattacks, are briefly surveyed.

1.3.1 Communication Delays

It is well recognized that communication delays are inherent in communications systems and have a detrimental impact on the control performance. For multi-agent systems, there are many results available in the literature concerning the effects of communications on distributed consensus algorithms [41–47]. In [42], the leader-following consensus of second-order multi-agent systems with coupling time delays under fixed and switching topologies was considered. Different from [42], tracking a varying-velocity leader with a time-varying delay was studied [43]. In [44], consensus problems for second-order agents with communication delay under directed and arbitrarily switching topologies were investigated. Considering the aperiodic sampling and network-induced delays, a distributed formation control protocol for linear multi-agent systems was proposed to achieve cluster formation [45]. In [46], synchronization of nonlinear multi-agent systems with time-varying control in the presence of transmission delays was studied. The containment control problem for the discrete-time first-order multi-agent systems was discussed and the necessary and sufficient condition for containment control with time delays was derived by using z-transformation and the Routh Criterion [47].

The effects of communication delays on the secondary frequency control of an islanded microgrid with multiple distributed generators were investigated [48]. Small-signal analysis of an islanded microgrid composed of two or more voltage-source inverters connected in parallel was conducted under a delayed communication network [49]. Droop-based distributed control algorithm considering the time-varying delays was proposed for microgrids [50]. In [51], two secondary control schemes were discussed to deal with the effects of delays. A small-signal dynamic model of microgrids equipped with time-delayed distributed secondary voltage control was developed [52, 53]. For AC microgrids, both output voltage/frequency restoration and active power sharing can be achieved by designing a distributed cooperative secondary control considering stochastic Markov time-varying network-induced delays [54]. Based on the Lyapunov-Krasovskii method, the effects of communication delays on the proposed novel secondary voltage and frequency protocols were characterized [55]. A distributed fixed-time multi-agent based control strategy with communication delays was presented for the frequency restoration, voltage regulation, state of charge balancing, and proportional reactive power sharing between photovoltaic battery systems distributed in a microgrid [56]. In [57], a resilient distributed H_∞ consensus-based control scheme was proposed in the secondary layer of AC heterogeneous autonomous microgrids, where fixed time-delays were taken into account.

By using an H_∞ delay-dependent control method, an issue of distributed coordination control for DC microgrids considering transmission delays was addressed [58]. Then, this work was extended to deal with the case of switching communication topology [59]. In [60], a distributed secondary control algorithm was proposed to realize voltage recovery and load sharing simultaneously, where constant or time-varying communication delays were taken into account, respectively. For DC microgrids with BESSs, a distributed multi-agent finite-time control strategy with different time delays was proposed to realize the state of charge balancing and voltage restoration [61]. A robust neighbor-based distributed cooperative control strategy was proposed for DC cyber-physical microgrids with communication delays and slow switching topologies [62].

1.3.2 Event-Triggered Communication Schemes

In conventional distributed control design, it is assumed that multi-agent systems have an ideal communication environment and can have access to high data flow through communication networks. However, in real implementation, communication resources such as communication bandwidth and channels are always limited and unlikely to support high communication data. Thus, an important criterion on designing a suitable distributed control scheme for multi-agent systems should not only guarantee the desirable control performance but also be able to make a full use of limited communication and computation resources. As a promising approach, event-triggered communication schemes are proposed to provide an effective solution to this issue [63–70]. It is verified that an event-triggered scheme is advantageous over a traditional time-triggered sampling scheme in reducing the utilization of communication and computation resources with the prescribed control performance [71, 72]. As a result, a great number of research results on distributed event-triggered consensus of multi-agent systems are reported from various perspectives [73–77], which are reviewed in [78, 79].

For networked microgrids, distributed control and optimization under event-triggered communications have received considerable attention and there are some results available in the literature [80–85]. For example, a distributed event-triggered algorithm was proposed to solve the economic dispatch problem for smart grids [80], where the initial balance of power supply and demand was required. In [82], distributed reactive power sharing for inverter-based microgrids under an event-triggered communication scheme was investigated. To fairly allocate total loads to different generators, a distributed event-triggered average consensus algorithm was introduced [83], where the initial values of the variables needed to be imposed on maximum capacity of generations and local load demands. However, the algorithm in [83] is not involved with the issue of frequency regulation. A distributed event-triggered cooperative control framework was presented to realize power sharing, frequency restoration and voltage control in inverter-based isolated microgrids [84]. In [85], distributed secondary control considering an event-triggered communication strategy was proposed to achieve frequency and voltage restoration, whereas the initial supply and demand balance was required. Furthermore, for DC microgrids, some works [86–90] were conducted to address both the issues of accurate current sharing and voltage regulation simultaneously by using distributed controllers with consideration of event-triggered communication.

1.3.3 Cyber Attacks

While adoption of communication networks facilitates distributed control strategies, it leads to vulnerability of systems to malicious cyber attacks that can be injected in a stealthy and unpredictable way. Cyber attacks are regarded as a serious threat in security and resilience of networked systems [91], since they can disrupt data transmission in medium channels by inserting some malware (e.g. viruses and worms) and destroy normal system operation by illegally invading monitoring centers. Roughly, cyber attacks can be categorized into: deception attacks and denial of service (DoS) attacks. The deception attack is to modify integrity of transmitted data packets while the DoS attack is to make data transmission unavailable by jamming shared network mediums.

Recently, a number of results [92–98] have been reported to investigate cyber security issues for microgrids. In [92], a distributed energy management algorithm was proposed in presence of misbehaving generation units caused by malicious cyber attacks. In [93], a cyber attack resilient distributed control strategy was developed against time-varying attack signals and successive attacks. A distributed observer-based finite-time secondary control scheme was proposed to improve the resilience of islanded AC microgrids under attacks [94]. In [95], a fully distributed resilient control framework was proposed to ensure system stability and preserve bounded synchronization under unbounded malicious attacks. An attack-resilient event-triggered controller was designed for a DC microgrid with multiple nonlinear constant power loads and intermittent DoS attacks [96]. A fully distributed attack-resilient secondary control framework was presented for DC microgrids, in the presence of unknown unbounded attacks on control input channels [97]. In [98], an attack detection strategy was proposed for DC microgrids by introducing a novel cooperative vulnerability factor framework for each agent to accurately identify the attacked units.

1.4 Book Outlines

The focus of this monograph is on multi-agent systems-based distributed coordinated control and optimization in networked micorgrids under various communication constraints. The monograph is organized as follows. In this chapter, some concepts and backgrounds for micorgrids are introduced, and recent developments on distributed control and optimization of micorgrids are briefly reviewed. In Chap. 2, network-based cooperative control for multi-agent systems is presented, where communication delays are taken into account. In Chap. 3, a sampled-data-based event-triggered scheme for distributed cooperative control in multi-agent systems is designed to reduce the usage of communication resources while maintaining the desired control performance. Chapter 4 deals with active power sharing and frequency regulation in an islanded micorgrid under event-triggered communication. In Chap. 5, secondary frequency and voltage control issues in islanded micorgrids with communication delays and switching topologies are investigated. Chapter 6 is concerned with distributed resilient finite-time secondary control for multiple heterogeneous BESSs in micorgrids subject to DoS attacks. In Chap. 7, a multi-objective optimization approach to realizing voltage regulation and load sharing for DC micorgrids is proposed, where the effects of communication delays are investigated. Chapter 8 addresses the issue of distributed energy management for smart grids under an event-triggered communication scheme.

References

1. A. Mohammed, S.S. Refaat, S. Bayhan, H. Abu-Rub, AC microgrid control and management strategies: evaluation and review. IEEE Power Electron. Mag. **6**(2), 18–31 (2019)
2. D.E. Olivares, A. Mehrizi-Sani, A.H. Etemadi, C.A. Canizares, R. Iravani, M. Kazerani, A.H. Hajimiragha, O. Gomis-Bellmunt, M. Saeedifard, R. Palma-Behnke, Trends in microgrid control. IEEE Trans. Smart Grid **5**(4), 1905–1919 (2014)
3. J.M. Guerrero, J.C. Vasquez, J. Matas, L.G. De Vicuña, M. Castilla, Hierarchical control of droop-controlled ac and dc microgrids- general approach toward standardization. IEEE Trans. Ind. Electron. **58**(1), 158–172 (2011)
4. J.P. Lopes, C. Moreira, A. Madureira, Defining control strategies for microgrids islanded operation. IEEE Trans. Power Syst. **21**(2), 916–924 (2006)
5. F. Chen, M. Chen, Q. Li, K. Meng, Y. Zheng, J.M. Guerrero, D. Abbott, Cost-based droop schemes for economic dispatch in islanded microgrids. IEEE Trans. Smart Grid **8**(1), 63–74 (2017)
6. Q.-C. Zhong, Robust droop controller for accurate proportional load sharing among inverters operated in parallel. IEEE Trans. Ind. Electron. **60**(4), 1281–1290 (2013)
7. J.W. Simpson-Porco, F. Drfler, F. Bullo, Voltage stabilization in microgrids via quadratic droop control. IEEE Trans. Autom. Control **62**(3), 1239–1253 (2017)
8. K. Tan, X. Peng, P.L. So, Y.C. Chu, M. Chen, Centralized control for parallel operation of distributed generation inverters in microgrids. IEEE Trans. Smart Grid **3**(4), 1977–1987 (2012)
9. J.W. Simpson-Porco, Q. Shafiee, F. Dörfler, J.C. Vasquez, J.M. Guerrero, F. Bullo, Secondary frequency and voltage control of islanded microgrids via distributed averaging. IEEE Trans. Ind. Electron. **62**(11), 7025–7038 (2015)
10. A. Bidram, A. Davoudi, F.L. Lewis, J.M. Guerrero, Distributed cooperative secondary control of microgrids using feedback linearization. IEEE Trans. Power Syst. **28**(3), 3462–3470 (2013)
11. J. Schiffer, T. Seel, J. Raisch, T. Sezi, Voltage stability and reactive power sharing in inverter-based microgrids with consensus-based distributed voltage control. IEEE Trans. Control Syst. Technol. **24**(1), 96–109 (2016)
12. Y. Xu, H. Sun, W. Gu, Y. Xu, Z. Li, Optimal distributed control for secondary frequency and voltage regulation in an islanded microgrid. IEEE Trans. Ind. Inform. **15**(1), 225–235 (2019)
13. X. Lu, X. Yu, J. Lai, J.M. Guerrero, H. Zhou, Distributed secondary voltage and frequency control for islanded microgrids with uncertain communication links. IEEE Trans. Ind. Inform. **13**(2), 448–460 (2017)

14. X. Lu, X. Yu, J. Lai, J.M. Guerrero, Y. Wang, H. Zhou, A novel distributed secondary coordination control approach for islanded microgrids. IEEE Trans. Smart Grid **9**(4), 2726–2740 (2018)
15. T. Morstyn, B. Hredzak, V.G. Agelidis, Distributed cooperative control of microgrid storage. IEEE Trans. Power Syst. **30**(5), 2780–2789 (2015)
16. J. Khazaei, Z. Miao, Consensus control for energy storage systems. IEEE Trans. Smart Grid **9**(4), 3009–3017 (2018)
17. D.H. Nguyen, J. Khazaei, Multiagent time-delayed fast consensus design for distributed battery energy storage systems. IEEE Trans. Sustain. Energy **9**(3), 1397–1406 (2018)
18. J. Khazaei, D.H. Nguyen, Multi-agent consensus design for heterogeneous energy storage devices with droop control in smart grids. IEEE Trans. Smart Grid **10**(2), 1395–1404 (2019)
19. T. Morstyn, B. Hredzak, V.G. Agelidis, Cooperative multi-agent control of heterogeneous storage devices distributed in a DC microgrid. IEEE Trans. Power Syst. **31**(4), 2974–2986 (2016)
20. J. Hu, A. Lanzon, Distributed finite-time consensus control for heterogeneous battery energy storage systems in droop-controlled microgrids. IEEE Trans. Smart Grid **10**(5), 4751–4761 (2019)
21. Z. Zuo, Q.-L. Han, B. Ning, X. Ge, X.-M. Zhang, An overview of recent advances in fixed-time cooperative control of multiagent systems. IEEE Trans. Ind. Inform. **14**(6), 2322–2334 (2018)
22. L. Meng, T. Dragicevic, J. Roldán-Pérez, J.C. Vasquez, J.M. Guerrero, Modeling and sensitivity study of consensus algorithm-based distributed hierarchical control for DC microgrids. IEEE Trans. Smart Grid **7**(3), 1504–1515 (2016)
23. S. Anand, B.G. Fernandes, J. Guerrero, Distributed control to ensure proportional load sharing and improve voltage regulation in low-voltage DC microgrids. IEEE Trans. Power Electron. **28**(4), 1900–1913 (2013)
24. V. Nasirian, S. Moayedi, A. Davoudi, F.L. Lewis, Distributed cooperative control of DC microgrids. IEEE Trans. Power Electron. **30**(4), 2288–2303 (2015)
25. Z. Zhang, M.-Y. Chow, Convergence analysis of the incremental cost consensus algorithm under different communication network topologies in a smart grid. IEEE Trans. Power Syst. **27**(4), 1761–1768 (2012)
26. S. Yang, S. Tan, J.-X. Xu, Consensus based approach for economic dispatch problem in a smart grid. IEEE Trans. Power Syst. **28**(4), 4416–4426 (2013)
27. G. Binetti, A. Davoudi, F.L. Lewis, D. Naso, B. Turchiano, Distributed consensus-based economic dispatch with transmission losses. IEEE Trans. Power Syst. **29**(4), 1711–1720 (2014)
28. N. Li, L. Chen, S. H. Low, Optimal demand response based on utility maximization in power networks, in *Proceedings of 2011 IEEE Power and Energy Society General Meeting*, Detroit, MI, 1–8 (2011)
29. P. Samadi, H. Mohsenian-Rad, R. Schober, V.W. Wong, Advanced demand side management for the future smart grid using mechanism design. IEEE Trans. Smart Grid **3**(3), 1170–1180 (2012)
30. Z. Fan, A distributed demand response algorithm and its application to PHEV charging in smart grids. IEEE Trans. Smart Grid **3**(3), 1280–1290 (2012)
31. N. Li, L. Chen, S. H. Low, Demand response in radial distribution networks: Distributed algorithm, in Proceedings of the 46th Asilomar Conference on Signal Systems and Computers, Pacific Grove, CA (2012), pp. 1549–1553
32. A. Safdarian, M. Fotuhi-Firuzabad, M. Lehtonen, A distributed algorithm for managing residential demand response in smart grids. IEEE Trans. Ind. Inform. **10**(4), 2385–2393 (2014)
33. E. Allen, M. Ilic, Z. Younes, Providing for transmission in times of scarcity: An ISO cannot do it all. Int. J. Elec. Power **21**(2), 147–163 (1999)
34. L. Wang, C. Wang, G. Yin, Y. Wang, Weighted and constrained consensus for distributed power flow control, in *Proceedings of 12th International Conference on Probabilistic Methods Applied to Power Systems PMAPS*, Istanbul, Turkey (2012)
35. L.Y. Wang, C. Wang, G. Yin, Y. Wang, Weighted and constrained consensus for distributed power dispatch of scalable microgrids. Asian J. Control **17**(5), 1725–1741 (2015)

36. H. Xing, Y. Mou, M. Fu, Z. Lin, Distributed bisection method for economic power dispatch in smart grid. IEEE Trans. Power Syst. **30**(6), 3024–3035 (2015)
37. G. Chen, J. Ren, E.N. Feng, Distributed finite-time economic dispatch of a network of energy resources. IEEE Trans. Smart Grid **8**(2), 822–832 (2017)
38. N. Rahbari-Asr, U. Ojha, Z. Zhang, M.-Y. Chow, Incremental welfare consensus algorithm for cooperative distributed generation/demand response in smart grid. IEEE Trans. Smart Grid **5**(6), 2836–2845 (2014)
39. C. Zhao, J. He, P. Cheng, J. Chen, Consensus-based energy management in smart grid with transmission losses and directed communication. IEEE Trans. Smart Grid **8**(5), 2049–2061 (2017)
40. Y. Xu, Z. Yang, W. Gu, M. Li, Z. Deng, Robust real-time distributed optimal control based energy management in a smart grid. IEEE Trans. Smart Grid **8**(4), 1568–1579 (2017)
41. R. Olfati-Saber, R. Murray, Consensus problems in networks of agents with switching topology and time-delays. IEEE Trans. Automat. Control **49**(9), 1520–1533 (2004)
42. J. Hu, Y. Hong, Leader-following coordination of multi-agent systems with coupling time delays. Phys. A **374**, 853–863 (2007)
43. K. Peng, Y. Yang, Leader-following consensus problem with a varying-velocity leader and time-varying delays. Phys. A **388**(2–3), 193–208 (2009)
44. J. Qin, H. Gao, W.X. Zheng, Second-order consensus for multi-agent systems with switching topology and communication delay. Syst. Control Lett. **60**(6), 390–397 (2011)
45. X. Ge, Q.-L. Han, X.-M. Zhang, Achieving cluster formation of multi-agent systems under aperiodic sampling and communication delay. IEEE Trans. Ind. Electron. **65**(4), 3417–3426 (2018)
46. Q. Jia, Z. Han, W. K. S. Tang, Synchronization of multi-agent systems with time-varying control and delayed communications. IEEE Trans. Circuits Syst. I: Regul. Pap. **66**(11), 4429–4438 (2019)
47. D. Wang, D. Wang, W. Wang, Necessary and sufficient conditions for containment control of multi-agent systems with time delay. Automatica **103**, 418–423 (2019)
48. S. Liu, X. Wang, P.X. Liu, Impact of communication delays on secondary frequency control in an islanded microgrid. IEEE Trans. Ind. Electron. **62**(4), 2021–2031 (2015)
49. E.A. Coelho, D. Wu, J.M. Guerrero, J.C. Vasquez, T. Dragicevi, C. Stefanovi, P. Popovski, Small-signal analysis of the microgrid secondary control considering a communication time delay. IEEE Trans. Ind. Electron. **63**(10), 6257–6269 (2016)
50. J. Lai, H. Zhou, X. Lu, X. Yu, W. Hu, Droop-based distributed cooperative control for microgrids with time-varying delays. IEEE Trans. Smart Grid **7**(4), 1775–1789 (2016)
51. C. Ahumada, R. Crdenas, D. Sez, J.M. Guerrero, Secondary control strategies for frequency restoration in islanded microgrids with consideration of communication delays. IEEE Trans. Smart Grid **7**(3), 1430–1441 (2016)
52. G. Lou, W. Gu, X. Lu, Y. Xu, H. Hong, Distributed secondary voltage control in islanded microgrids with consideration of communication network and time delays. IEEE Trans. Smart Grid **11**(5), 3702–3715 (2020)
53. G. Lou, W. Gu, Y. Xu, W. Jin, X. Du, Stability robustness for secondary voltage control in autonomous microgrids with consideration of communication delays. IEEE Trans. Power Syst. **33**(4), 4164–4178 (2018)
54. C. Zhao, W. Sun, J. Wang, Q. Li, D. Mu, X. Xu, Distributed cooperative secondary control for islanded microgrid with markov time-varying delays. IEEE Trans. Energy Conver. **34**(4), 2235–2247 (2019)
55. G. Chen, Z. Guo, Distributed secondary and optimal active power sharing control for islanded microgrids with communication delays. IEEE Trans. Smart Grid **10**(2), 2002–2014 (2019)
56. R. Zhang, B. Hredzak, Nonlinear Sliding Mode and Distributed control of battery energy storage and photovoltaic systems in AC microgrids with communication delays. IEEE Trans. Ind. Inform. **15**(9), 5149–5160 (2019)
57. M. Raeispour, H. Atrianfar, H. Baghaee, G.B. Gharehpetian, Resilient H_∞ consensus-based control of autonomous AC microgrids with uncertain time-delayed communications. IEEE Trans. Smart Grid **11**(5), 3871–3884 (2020)

58. C. Dou, D. Yue, J.M. Guerrero, X. Xie, S. Hu, Multiagent system-based distributed coordinated control for radial DC microgrid considering transmission time delays. IEEE Trans. Smart Grid **8**(5), 2370–2381 (2017)
59. C. Dou, D. Yue, Z. Zhang, K. Ma, MAS-Based distributed cooperative control for DC microgrid through switching topology communication network with time-varying delays. IEEE Syst. J. **13**(1), 615–624 (2019)
60. M. Dong, L. Li, Y. Nie, D. Song, J. Yang, Stability analysis of a novel distributed secondary control considering communication delay in DC microgrids. IEEE Trans. Smart Grid **10**(6), 6690–6700 (2019)
61. R. Zhang, B. Hredzak, Distributed Finite-Time multiagent montrol for DC microgrids with time felays. IEEE Trans. Smart Grid **10**(3), 2692–2701 (2019)
62. J. Lai, X. Lu, X. Yu, A. Monti, H. Zhou, Distributed voltage regulation for cyber-physical microgrids with coupling delays and slow switching topologies. IEEE Trans. Syst. Man Cyben. Syst. **50**(1), 100–110 (2020)
63. P. Tabuada, Event-triggered real-time scheduling of stabilizing control tasks. IEEE Trans. Autom. Control. **52**(9), 1680–1685 (2007)
64. X. Wang, M. Lemmon, Event-triggering in distributed networked control systems. IEEE Trans. Autom. Control **56**(3), 586–601 (2011)
65. M. Donkers, W. Heemels, Output-based event-triggered control with guaranteed L_∞-gain and improved and decentralized event-triggering. IEEE Trans. Autom. Control **57**(6), 1362 (2012)
66. C. Peng, Q.-L. Han, A novel event-triggered transmission scheme and $\mathscr{L}_2$ control co-design for sampled-data control systems. IEEE Trans. Autom. Control **58**(10), 2620–2626 (2013)
67. W. Heemels, M. Donkers, A. Teel, Periodic event-triggered control for linear systems. IEEE Trans. Autom. Control **58**(4), 847–861 (2013)
68. D. Yue, E. Tian, Q.-L. Han, A delay system method for designing event-triggered controllers of networked control systems. IEEE Trans. Autom. Control **58**(2), 475–481 (2013)
69. X.-M. Zhang, Q.-L. Han, A decentralized event-triggered dissipative control scheme for systems with multiple sensors to sample the system outputs. IEEE Trans. Cybern. **46**(12), 2745–2757 (2016)
70. B.-L. Zhang, Q.-L. Han, X.-M. Zhang, Event-triggered H_∞ reliable control for offshore structures in network environments. J. Sound Vib. **368**, 1–21 (2016)
71. X.-M. Zhang, Q.-L. Han, X. Yu, Survey on recent advances in networked control systems. IEEE Trans. Ind. Inform. **12**(5), 1740–1752 (2016)
72. X.-M. Zhang, Q.-L. Han, B.-L. Zhang, An overview and deep investigation on sampled-data-based event-triggered control and filtering for networked systems. IEEE Trans. Ind. Inform. **13**(1), 4–16 (2017)
73. G. Seyboth, D. Dimarogonas, K. Johansson, Event-based broadcasting for multi-agent average consensus. Automatica **49**, 245–252 (2013)
74. Y. Fan, G. Feng, Y. Wang, C. Song, Distributed event-triggered control of multi-agent systems with combinational measurements. Automatica **49**(2), 671–675 (2013)
75. W. Zhu, Z.-P. Jiang, G. Feng, Event-based consensus of multi-agent systems with general linear models. Automatica **50**(2), 552–558 (2014)
76. C. Nowzari, J. Cortés, Distributed event-triggered coordination for average consensus on weight-balanced digraphs. Automatica **68**, 237–244 (2016)
77. E. Garcia, Y. Cao, D. Casbeer, Decentralized event-triggered consensus with general linear dynamics. Automatica **50**(10), 2633–2640 (2014)
78. X. Ge, F. Yang, Q.-L. Han, Distributed networked control systems: A brief overview. Inform. Sci. **380**, 117–131 (2017)
79. J. Qin, Q. Ma, Y. Shi, L. Wang, Recent advances in consensus of multi-agent systems: A brief survey. IEEE Trans. Ind. Electron. **64**(6), 4972–4983 (2017)
80. C. Li, X. Yu, W. Yu, T. Huang, Z.-W. Liu, Distributed event-triggered scheme for economic dispatch in smart grids. IEEE Trans. Ind. Inform. **12**(5), 1775–1785 (2016)
81. M. Tahir, S.K. Mazumder, Self-triggered communication enabled control of distributed generation in microgrids. IEEE Trans. Ind. Inform. **11**(2), 441–449 (2015)

82. Y. Fan, G. Hu, M. Egerstedt, Distributed reactive power sharing control for microgrids with event-triggered communication. IEEE Trans. Control Syst. Technol. **25**(1), 118–128 (2017)
83. W. Meng, X. Wang, S. Liu, Distributed load sharing of an inverter-based microgrid with reduced communication. IEEE Trans. Smart Grid **9**(2), 1354–1364 (2018)
84. S. Weng, D. Yue, C. Dou, J. Shi, C. Huang, Distributed event-triggered cooperative control for frequency and voltage stability and power sharing in isolated inverter-based microgrid. IEEE Trans. Cybern. **49**(4), 1427–1439 (2018)
85. M. Chen, X. Xiao, J.M. Guerrero, Secondary restoration control of islanded microgrids with decentralized event-triggered strategy. IEEE Trans. Ind. Inform. **14**(9), 3870–3880 (2018)
86. R. Han, L. Meng, J.M. Guerrero, J.C. Vasquez, Distributed nonlinear control with event-triggered communication to achieve current-sharing and voltage regulation in DC Microgrids. IEEE Trans. Power Electron. **33**(7), 6416–6433 (2018)
87. F. Guo, L. Wang, C. Wen, D. Zhang, Q. Xu, Distributed voltage restoration and current sharing control in islanded DC microgrid systems without continuous communication. IEEE Trans. Ind. Electron. **67**(4), 3043–3053 (2020)
88. S. Sahoo, S. Mishra, An adaptive event-triggered communication-based distributed secondary control for DC microgrids. IEEE Trans. Smart Grid **9**(6), 6674–6683 (2018)
89. B. Fan, J. Peng, Q. Yang, W. Liu, Distributed periodic event-triggered algorithm for current sharing and voltage regulation in DC microgrids. IEEE Trans. Smart Grid **11**(1), 577–589 (2020)
90. J. Zhou, Y. Xu, H. Sun, L. Wang, M.-Y. Chow, Distributed event-triggered H_∞ consensus based current sharing control of DC microgrids considering uncertainties. IEEE Trans. Ind. Inform. **16**(12), 7413–7425 (2020)
91. D. Ding, Q.-L. Han, Y. Xiang, X. Ge, X.-M. Zhang, A survey on security control and attack detection for industrial cyber-physical systems. Neurocomputing **275**, 1674–1683 (2018)
92. W. Zeng, Y. Zhang, M.-Y. Chow, Resilient distributed energy management subject to unexpected misbehaving generation units. IEEE Trans. Ind. Inform. **13**(1), 208–216 (2017)
93. Q. Zhou, M. Shahidehpour, A. Alabdulwahab, A. Abusorrah, A cyber-attack resilient distributed control strategy in islanded microgrids. IEEE Trans. Smart Grid **11**(5), 3690–3701 (2020)
94. R. Lu, J. Wang, Z. Wang, Distributed observer-based finite-time control of AC microgrid under attack. IEEE Trans. Smart Grid **12**(1), 157–168 (2021)
95. S. Zuo, O.A. Beg, F.L. Lewis, A. Davoudi, Resilient networked AC microgrids under unbounded cyber attacks. IEEE Trans. Smart Grid **11**(5), 3850–3859 (2020)
96. S. Hu, P. Yuan, D. Yue, C. Dou, Z. Cheng, Y. Zhang, Attack-resilient event-triggered controller design of DC microgrids under DoS attacks. IEEE Trans. Circuits Syst. I: Regul. Pap. **67**(2), 699–710(2020)
97. S. Zuo, T. Altun, F.L. Lewis, A. Davoudi, Distributed resilient secondary control of DC microgrids against unbounded attacks. IEEE Trans. Smart Grid **11**(5), 3785–3794 (2020)
98. S. Sahoo, S. Mishra, J. C.-H. Peng, T. Dragičević, A stealth cyber-attack detection strategy for DC microgrids. IEEE Trans. Power Electron. **34**(8), 8162–8174 (2019)

Chapter 2
Network-Based Consensus of Multi-agent Systems

This chapter is concerned with network-based distributed leader-following consensus for multi-agent systems. A network-based consensus control protocol under a directed graph is proposed, with which each agent can be remotely controlled via a communication network. A delay-dependent stability criterion for the multi-agent system is derived by constructing a Lyapunov-Krasovskii functional with graph information. Then, by employing this stability criterion, a delay-dependent sufficient condition for the existence of network-based consensus controllers is formulated in terms of linear matrix inequalities, which ensures that all the follower agents' states can reach an agreement on the leader's state. A numerical example is given to illustrate the effectiveness of the obtained results.

2.1 Graph Theory

Let $\mathscr{G} = (\mathscr{V}, E)$ denote a directed weighted graph, where $\mathscr{V} = \{1, 2, \ldots, N\}$ is the set of nodes and $E \subseteq \mathscr{V} \times \mathscr{V}$ is the set of edges. Node j is a neighbor of node i if there is an edge defined as $\varepsilon_{ij} \triangleq (i, j) \in E$. $\mathscr{N}_i \triangleq \{j |\ \varepsilon_{ij} \in E\}$ represents the set of neighbors of node i. In contrast to a directed graph, the pairs of nodes in an undirected graph are unordered, where the edge (i, j) denotes that nodes i and j can obtain information from each other. Note that an undirected graph can be viewed as a special case of a directed graph. $\mathscr{A} = [a_{ij}] \in \mathbb{R}^{N\times N}$ is a weighted adjacency matrix, where $a_{ii} = 0$ for all i, and $a_{ij} > 0$ if $\varepsilon_{ij} \in E$, otherwise, $a_{ij} = 0$. The degree matrix of graph $\mathscr{G}$ is $\mathscr{D} = \text{diag}\{d_1, d_2, \ldots, d_N\}$, where the diagonal elements are represented as $d_i = \sum_{j\in\mathscr{N}_i} a_{ij}$. The Laplacian matrix L of the graph is defined as $L = \mathscr{D} - \mathscr{A}$.

L. Ding et al., *Distributed Control and Optimization of Networked Microgrids*, Power Systems, https://doi.org/10.1007/978-3-030-95029-3_2

A directed path from node i to node j is a sequence of edges of the form (i, i_2), (i_2, i_3),... (i_k, j) in a directed graph. A directed tree is a directed graph in which every node has exactly one parent except for one node, called the root, which has no parent and which has a directed path to every other node. A directed spanning tree of a directed graph is a directed tree that contains all nodes of the directed graph. A directed graph contains a directed spanning tree if there exists a directed spanning tree as a subset of the directed graph. The undirected graph $\mathcal{G}$ is connected if there exists a path between any nodes. The matrix L of a connected graph is symmetric and has a simple zero eigenvalue and all other eigenvalues are real and positive [1].

Denote $\tilde{\mathcal{G}}$ as a graph which contains N follower nodes and a leader node. Similarly, a diagonal matrix $M = diag\{m_1, m_2, \ldots, m_n\} \in \mathbb{R}^{N\times N}$ is referred to as the leader adjacency matrix with $m_i \geq 0$ for any i. If the leader is a neighbor of node v_i, then $m_i > 0$; otherwise, $m_i = 0$. If there is a path in $\tilde{\mathcal{G}}$ from every node i in $\mathcal{G}$ to node 0, then node 0 is globally reachable in $\tilde{\mathcal{G}}$.

2.2 Problem Statement

Consider a distributed multi-agent system consisting of N follower agents and a leader. The dynamics of each follower agent can be described by

$$\begin{cases} \dot{x}_i = Ax_i(t) + Bu_i(t), \\ x_i(t_0) = x_i^0, \quad i = 1, 2, \ldots, N, \end{cases} \tag{2.1}$$

where $A \in \mathbb{R}^{n\times n}$ and $B \in \mathbb{R}^{n\times m}$ are constant matrices; $x_i \in \mathbb{R}^n$ and $u_i \in \mathbb{R}^m$ are the state and the control input of agent i, respectively. The leader is labeled as $i = 0$ and its dynamics can be expressed as

$$\dot{x}_0(t) = Ax_0(t), \qquad x_0(t_0) = x_0^0, \tag{2.2}$$

where $x_0 \in \mathbb{R}^n$ represents the state of the leader agent.

We need the following necessary assumptions.

Assumption 2.2.1 The pair (A, B) is stabilizable.

Assumption 2.2.2 The leader node 0 is globally reachable in Graph $\tilde{\mathcal{G}}$. □

The control objective is to design an appropriate network-based feedback control $u_i(t)$ such that all the follower agents' states asymptotically reach consensus with the leader's state, i.e.

$$\lim_{t\to\infty} \|x_i(t) - x_0(t)\| = 0 \tag{2.3}$$

for any initial condition $x_i(t_0)$ and $i = 1, 2, \ldots, N$.

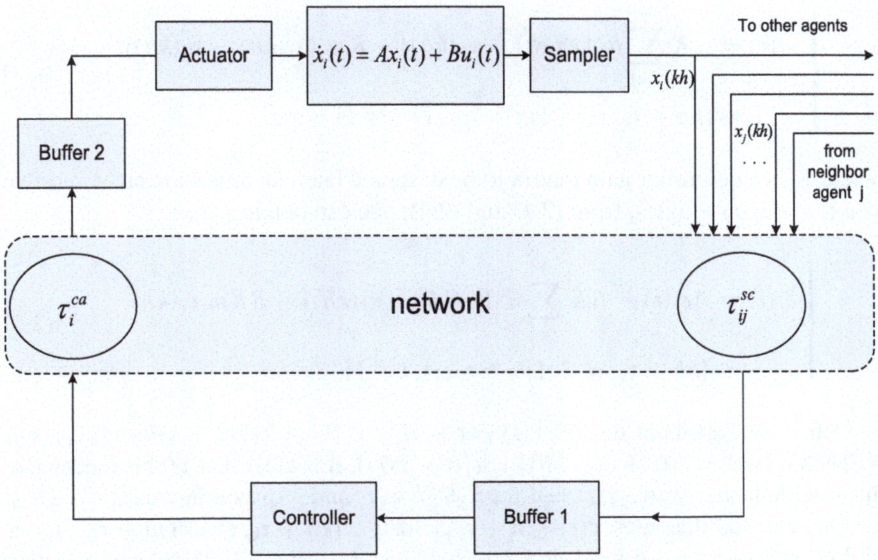

Fig. 2.1 Network-based consensus feedback control for agent i

In this chapter, each agent in the distributed multi-agent system is controlled by a network-based consensus feedback controller, which is shown in Fig. 2.1. It is assumed that the sampler is clock-driven and all the controller and the buffers are event-driven. Each agent's information is sampled by a sampler at a period $h > 0$. The sampled-data of agent i can be successfully sent to the controller and the other agents, and then to the actuator via a communication network. Considering the effects of network-induced delays, the sampled-data from agent i and its neighbor j may arrive at the controller of agent i at different times. Thus, Buffer 1 is employed to store the received data information until all the information transmission for agent i is completed, and then all these data information will be sent to the controller for computing again. Let τ_{ij}^{sc} for $i = 1 \ldots, N$, $j = 0, 1 \ldots, N$ be the communication delay of agent i from all the samplers of the agent's own and its neighbor agents j to Buffer 1. As a result, the time when all the data for agent i arrives at the controller can be represented as $\tau_i^{sc} = \max_j\{\tau_{ij}^{sc} \mid j = 0, 1 \ldots, N\}$ for $i = 1 \ldots, N$. In order to make all the agents be evolved in the same time scale, Buffer 2 is used to store data information available until all the agents completely achieve the kth sampled information exchange between the controller and Buffer 2. Let τ_i^{ca} be the communication delay of agent i from the controller to Buffer 2. So the time delay $\tau_{ik} = \tau_{ik}^{sc} + \tau_{ik}^{ca}$ for $i = 1 \ldots, N$ denotes the time when data information at the kth sampling time from the sampler of agent i and its neighbor agent j is successfully transmitted to Buffer 2. Consequently, the total time when the kth sampled-data of all the agents in the distributed multi-agent system is successfully transmitted from the samplers to the corresponding agent's actuators can be defined as $\tau_k = \max_i\{\tau_{ik} \mid i = 1, \ldots, N\}$. Then, the consensus control protocol for system (2.1) is designed as

$$\begin{cases} u_i(t) = -K\sum_{j=1}^{N} a_{ij}[x_i(kh) - x_j(kh)] - Km_i[x_i(kh) - x_0(kh)], \\ t \in [kh + \tau_k, (k+1)h + \tau_{k+1}), \ \ k \in \mathbb{N}, \end{cases} \tag{2.4}$$

where K is a controller gain matrix to be designed later. Defining the error variable as $e_i(t) = x_i(t) - x_0(t)$, from (2.1) and (2.4), one can obtain

$$\begin{cases} \dot{e}_i(t) = Ae_i(t) - BK\sum_{j=1}^{N} a_{ij}[e_i(kh) - e_j(kh)] - BKm_ie_i(kh), \\ t \in [kh + \tau_k, (k+1)h + \tau_{k+1}), k \in \mathbb{N}. \end{cases} \tag{2.5}$$

Define an "artificial delay" $\tau(t) = t - kh, t \in [kh + \tau_k, (k+1)h + \tau_{k+1}), k \in \mathbb{N}$. Then $e_i(kh) = e_i(t - (t - kh)) = e_i(t - \tau(t))$, It is clear that $\tau(t)$ is piecewise-linear with the derivative $\dot{\tau}(t) = 1$ for $t \neq kh + \tau_k$ and is discontinuous at $t = kh + \tau_k$. One can see that $\tau_k \leq \tau(t) \leq h + \tau_{k+1}$ for $t \in [kh + \tau_k, (k+1)h + \tau_{k+1}), k \in \mathbb{N}$. Let $\alpha = \min_k\{\tau_k \mid k \in \mathbb{N}\}$ and $\beta = h + \max_k\{\tau_k \mid k \in \mathbb{N}\}$. Then, we have that $0 \leq \alpha \leq \tau(t) \leq \beta < +\infty$.

Next, defining $e(t) = [e_1^T(t), e_2^T(t), \ldots, e_N^T(t)]^T$ and along with (2.5), we have

$$\begin{cases} \dot{e}(t) = (I_N \otimes A)e(t) - (H \otimes BK)(e(t - \tau(t)), \\ t \in [kh + \tau_k, (k+1)h + \tau_{k+1}), k \in \mathbb{N}, \end{cases} \tag{2.6}$$

where the matrix $H = L + M$ and the sign $\otimes$ represents matrix Kronecker product. The initial condition of the state $e(t)$ is supplemented as

$$e(\theta) = \phi(\theta), \quad \theta \in [t_0 - \beta, t_0] \tag{2.7}$$

where $\phi(t_0) = e(t_0) = [e_1^T(t_0), e_2^T(t_0), \ldots, e_N^T(t_0)]^T$. Let $e_t(\theta) = e(t + \theta)$, $\theta \in [t_0 - \beta, t_0]$, $\phi(\theta) = e_t(\theta)$ and $\phi \in W$, where W denotes the Banach space of absolutely continuous functions $[t_0 - \beta, t_0] \to \mathbb{R}^{Nn}$ with square-integrable derivatives and the norm

$$\|\phi\|_W = \max_{\theta \in [t_0 - \beta, t_0]} \|\phi(\theta)\| + \left[\int_{t_0 - \beta}^{t_0} \|\dot{\phi}(s)\|^2 ds\right]^{\frac{1}{2}}. \tag{2.8}$$

Remark 2.1 It should be pointed out that data packet dropouts and disorders may happen in the process of information transmission. In this chapter, however, in order to simplify the analysis, we only consider the effects of network-induced delays.

2.3 Stability Analysis

In this section, we will derive a stability criterion for the system (2.6)–(2.7). For this purpose, the following lemmas are needed.

Lemma 2.1 *Let L_{ug} denote the Laplacian matrix of the undirected graph $\bar{\mathscr{G}}$ associated with $\mathscr{G}$. If node 0 is globally reachable in graph $\tilde{\mathscr{G}}$, then the symmetric matrix $U = L_{ug} + M$ is positive definite.*

Proof Note that L_{ug} is symmetric. Let $\bar{\mathscr{G}}$ be an undirected graph associated with $\tilde{\mathscr{G}}$. If node 0 is globally reachable in graph $\tilde{\mathscr{G}}$, then graph $\bar{\mathscr{G}}$ is connected. By Lemma 4 in [2], one can conclude that U is positive definite. □

Lemma 2.2 *Suppose that there exist scalars $\varepsilon_i > 0$ $(i = 1, 2, 3)$ and a Lyapunov functional $V(t, e_t, \dot{e}_t) : \mathbb{R} \times \mathbb{R}^{Nn} \times \mathbb{R}^{Nn} \to \mathbb{R}$ such that the corresponding solution e_t to system (2.6) is defined for $t \geq t_0$, and $V(t, e_t, \dot{e}_t)$ is continuously differentiable at $t \neq kh + \tau_k$ and satisfies*

$$\varepsilon_1 \|e(t)\|^2 \leq V(t, e_t, \dot{e}_t) \leq \varepsilon_2 \|e_t\|_W^2, \tag{2.9}$$

$$\frac{dV(t, e_t, \dot{e}_t)}{dt} \leq -\varepsilon_3 \|e(t)\|^2, \quad t \neq kh + \tau_k, \tag{2.10}$$

$$\lim_{t \uparrow (kh+\tau_k)} V(t, e_t, \dot{e}_t) \geq V(t, e_t, \dot{e}_t) \,|_{t=kh+\tau_k} \tag{2.11}$$

For briefness of representation, let $\psi(t) = [e^T(t)\; e^T(t - \tau(t))\; e^T(t - \alpha)\; e^T(t - \beta)\; e^T(t - \rho(t))]^T$ and $\mathscr{E}_i (i = 1, 2, \ldots, 5)$ be a block entry matrix with $\mathscr{E}_{ij} = \mathscr{E}_i - \mathscr{E}_j$. For example, $\mathscr{E}_2 = [0, I_{Nn}, 0, 0, 0]$ and $\mathscr{E}_{24} = [0, I_{Nn}, 0, -I_{Nn}, 0]$.

We construct a Lyapunov-Krasovskii functional candidate as

$$V(t, e_t, \dot{e}_t) = V_1(t, e_t, \dot{e}_t) + V_2(t, e_t, \dot{e}_t), \tag{2.12}$$

where

$$\begin{aligned} V_1(t, e_t, \dot{e}_t) &= e^T(t)(U \otimes P)e(t) + \int_{t-\alpha}^{t} e^T(s)(U \otimes Q_1)e(s)ds \\ &\quad + \int_{t-\beta}^{t-\alpha} e^T(s)(U \otimes Q_2)e(s)ds \\ &\quad + \alpha \int_{-\alpha}^{0} \int_{t+\theta}^{t} \dot{e}^T(s)(U \otimes R_1)\dot{e}(s)d\theta ds \\ &\quad + \int_{-\beta}^{-\alpha} \int_{t+\theta}^{t} \dot{e}^T(s)(U \otimes R_2)\dot{e}(s)d\theta ds \\ V_2(t, e_t, \dot{e}_t) &= (\beta - \tau(t))\{[e(t) - e(t - \rho(t))]^T (U \otimes X) \\ &\quad \times [e(t) - e(t - \rho(t))] + \int_{t-\rho(t)}^{t} \dot{e}^T(s)(U \otimes Y)\dot{e}(s)ds\} \end{aligned}$$

with $\rho(t) = \tau(t) - \tau_k, t \in [kh + \tau_k, (k+1)h + \tau_{k+1}), k \in \mathbb{N}, P > 0, Q_l > 0, R_l > 0$ $(l = 1, 2)$, $X > 0$ and $Y > 0$. By Lemma 2.1 and the properties of Kronecker product, we have that $U \otimes P > 0, U \otimes Q_l > 0, U \otimes R_l > 0\,(l = 1, 2), U \otimes X > 0$, and $U \otimes Y > 0$.

Remark 2.2 It is well known that the direct Lyapunov method is a powerful tool for studying the stability of systems with time-delays. Note that in [3] a Lyapunov-Krasovskii functional was employed to consider a leader-following consensus problem for a multi-agent system with a varying-velocity leader and time-varying delays. This Lyapunov-Krasovskii functional unexpectedly made all the agents connected together even though two agents were relatively independent. As a result, some redundant information may be correspondingly introduced. However, in this chapter, we construct a Lyapunov-Krasovskii functional (2.12), which has two advantage aspects: (i) introducing the matrix U to avoid some unnecessary correlations between any independent agents; and (ii) making full use of the piecewise-linear sawtooth structure characteristic of $\tau(t)$ by introducing $V_2(t, e_t, \dot{e}_t)$.

Before giving our main theorem, the following lemmas are needed.

Lemma 2.3 ([4]) *For any constant matrix* $R > 0 \in \mathbb{R}^{n\times n}$, *a scalar* $\gamma > 0$, *and a vector function* $\dot{e} : [-\gamma, 0] \to \mathbb{R}^n$ *such that the following integration is well defined, then*

$$-\gamma \int_{t-\gamma}^{t} \dot{e}^T(s) R \dot{e}(s) ds \le -[e(t) - e(t-\gamma)]^T R [e(t) - e(t-\gamma)] \qquad (2.13)$$

Lemma 2.4 ([5]) *For any constant matrix* $R > 0 \in \mathbb{R}^{n\times n}$, *a scalar function* $\gamma(t)$ *with* $0 < \gamma(t) \le \gamma_M$, *and a vector function* $\dot{e} : [-\gamma_M, 0] \to \mathbb{R}^n$ *such that the following integration is well defined, let* $\int_{t-\gamma(t)}^{t} \dot{e}(s) ds = E\zeta(t)$ *where* $E \in \mathbb{R}^{n\times k}$ *and* $\zeta(t) \in \mathbb{R}^k$. *Then the following inequality holds for any matrix* $Z \in \mathbb{R}^{n\times k}$

$$-\int_{t-\gamma(t)}^{t} \dot{e}^T(s) R \dot{e}(s) ds \le \zeta^T(t)(\gamma(t) Z^T R^{-1} Z - E^T Z - Z^T E)\zeta(t) \qquad (2.14)$$

We now state and establish the following result.

Theorem 2.1 *Under Assumptions 2.2.1 and 2.2.2, given some scalars* α *and* β *with* $\beta \ge \alpha$, *the system described by* (2.6)–(2.7) *is asymptotically stable if there exist real matrices* $P > 0$, $Q_1 > 0$, $Q_2 > 0$, $R_1 > 0$, $R_2 > 0$, $X > 0$, $Y > 0$, *and matrices* Z_k $(k = 1, 2, 3)$ *of appropriate dimensions such that*

$$\begin{bmatrix} \Pi_{11}+\sigma\Upsilon_{11} & \Pi_{12}+\sigma\Upsilon_{12} & \sigma Z_1^T \\ * & \Pi_{22}+\Upsilon_{22} & 0 \\ * & * & -\sigma U\otimes R_2 \end{bmatrix} < 0 \tag{2.15}$$

$$\begin{bmatrix} \Pi_{11} & \Pi_{12} & \sigma Z_2^T & \sigma Z_3^T \\ * & \Pi_{22} & 0 & 0 \\ * & * & -\sigma U\otimes R_2 & 0 \\ * & * & * & -\sigma U\otimes Y \end{bmatrix} < 0 \tag{2.16}$$

where $\sigma = \beta - \alpha$ *and*

$$\begin{aligned}
\Pi_{11} =& \mathscr{E}_1^T[(U\otimes PA)\mathscr{E}_1 - (UH\otimes PBK)\mathscr{E}_2] \\
&+ [(U\otimes PA)\mathscr{E}_1 - (UH\otimes PBK)\mathscr{E}_2]^T\mathscr{E}_1 \\
&+ \mathscr{E}_1^T(U\otimes Q_1)\mathscr{E}_1 + \mathscr{E}_3^T(U\otimes(Q_2-Q_1))\mathscr{E}_3 \\
&- \mathscr{E}_4^T(U\otimes Q_2)\mathscr{E}_4 - \mathscr{E}_{13}^T(U\otimes R_1)\mathscr{E}_{13} \\
&- \mathscr{E}_{15}^T(U\otimes X)\mathscr{E}_{15} - \mathscr{E}_{24}^T Z_1 - Z_1^T\mathscr{E}_{24} - \mathscr{E}_{32}^T Z_2 \\
&- Z_2^T\mathscr{E}_{32} - \mathscr{E}_{15}^T Z_3 - Z_3^T\mathscr{E}_{15}, \\
\Pi_{12} =& [I_N\otimes A)\mathscr{E}_1 - (H\otimes BK)\mathscr{E}_2]^T[\alpha^2 U\otimes R_1 + \sigma U\otimes R_2], \\
\Upsilon_{11} =& \mathscr{E}_{15}^T[(U\otimes XA)\mathscr{E}_1 - (UH\otimes XBK)\mathscr{E}_2] \\
&+ [(U\otimes XA)\mathscr{E}_1 - (UH\otimes XBK)\mathscr{E}_2]^T\mathscr{E}_{15}, \\
\Upsilon_{12} =& [(U\otimes YA)\mathscr{E}_1 - (UH\otimes YBK)\mathscr{E}_2]^T, \\
\Pi_{22} =& U\otimes[-\alpha^2 R_1 - \sigma R_2], \\
\Upsilon_{22} =& -\sigma U\otimes Y.
\end{aligned}$$

Proof Taking the derivative of $V(t, e_t, \dot{e}_t)$ with respect to $t \in [kh+\tau_k, (k+1)h+\tau_{k+1})$, $k\in\mathbb{N}$ along the trajectory (2.6) yields

$$\begin{aligned}
\dot{V}(t, e_t, \dot{e}_t) =& \psi^T(t)\{2\mathscr{E}_1^T(U\otimes P)[(I_N\otimes A)\mathscr{E}_1 - (H\otimes BK)\mathscr{E}_2] \\
&+ \mathscr{E}_1^T(U\otimes Q_1)\mathscr{E}_1 - \mathscr{E}_4^T(U\otimes Q_2)\mathscr{E}_4 + \mathscr{E}_3^T(U\otimes(Q_2-Q_1))\mathscr{E}_3 \\
&+ [I_N\otimes A)\mathscr{E}_1 - (H\otimes BK)\mathscr{E}_2]^T(\alpha^2 U\otimes R_1 + \sigma U\otimes R_2) \\
&\times [(I_N\otimes A)\mathscr{E}_1 - (H\otimes BK)\mathscr{E}_2] + 2(\beta - \tau(t))\mathscr{E}_{15}^T[(U\otimes XA)\mathscr{E}_1 \\
&- (UH\otimes XBK)\mathscr{E}_2] - \mathscr{E}_{15}^T(U\otimes X)\mathscr{E}_{15} + (\beta - \tau(t)) \\
&\times [(I_N\otimes A)\mathscr{E}_1 - (H\otimes BK)\mathscr{E}_2]^T(U\otimes Y)[(I_N\otimes A)\mathscr{E}_1 \\
&- (H\otimes BK)\mathscr{E}_2]\}\psi(t) + \eta_1(t) + \eta_2(t) + \eta_3(t)
\end{aligned} \tag{2.17}$$

with

$$\eta_1(t) = -\alpha \int_{t-\alpha}^{t} \dot{e}^T(s)(U \otimes R_1)\dot{e}(s)ds$$
$$\eta_2(t) = -\int_{t-\beta}^{t-\alpha} \dot{e}^T(s)(U \otimes R_2)\dot{e}(s)ds$$
$$\eta_3(t) = -\int_{t-\rho(t)}^{t} \dot{e}^T(s)(U \otimes Y)\dot{e}(s)ds.$$

By using Lemma 2.3, we have

$$\eta_1(t) \leq -\psi^T(t)\mathscr{E}_{13}^T(U \otimes R_1)\mathscr{E}_{13}\psi(t) \tag{2.18}$$

Note that

$$\begin{aligned}
\eta_2(t) + \eta_3(t) = & -\int_{t-\beta}^{t-\tau(t)} \dot{e}^T(s)(U \otimes R_2)\dot{e}(s)ds \\
& -\int_{t-\tau(t)}^{t-\alpha} \dot{e}^T(s)(U \otimes R_2)\dot{e}(s)ds + \eta_3(t) \\
\leq & \, \psi^T(t)[(\beta - \tau(t))Z_1^T(U \otimes R_2)^{-1}Z_1 - \mathscr{E}_{24}^T Z_1 - Z_1^T \mathscr{E}_{24} \\
& + (\tau(t) - \alpha)Z_2^T(U \otimes R_2)^{-1}Z_2 - \mathscr{E}_{32}^T Z_2 - Z_2^T \mathscr{E}_{32} \\
& + (\tau(t) - \alpha)Z_3^T(U \otimes Y)^{-1}Z_3 - \mathscr{E}_{15}^T Z_3 - Z_3^T \mathscr{E}_{15}]\psi(t)
\end{aligned} \tag{2.19}$$

where Lemma 2.4 and $\rho(t) = \tau(t) - \tau_k \leq \tau(t) - \alpha$ are used for the last inequality. From (2.17)–(2.19), we obtain

$$\dot{V}(t, e_t, \dot{e}_t) \leq \psi^T(t)\Sigma\psi(t) \tag{2.20}$$

where

$$\begin{aligned}
\Sigma &= \Pi_{11} + \Omega_1 + (\beta - \tau(t))(\Upsilon_{11} + \Omega_2) + (\tau(t) - \alpha)\Omega_3 \\
\Omega_1 &= \Pi_{12}\Pi_{22}^{-1}\Pi_{12}^T \\
\Omega_2 &= \Upsilon_{12}\Upsilon_{22}^{-1}\Upsilon_{12}^T + Z_1^T(U \otimes R_2)^{-1}Z_1 \\
\Omega_3 &= Z_2^T(U \otimes R_2)^{-1}Z_2 + Z_3^T(U \otimes Y)^{-1}Z_3
\end{aligned}$$

It is clear that if $\Sigma < 0$, then there exists a scalar $\varepsilon_3 > 0$ such that $\dot{V}(t, e_t, \dot{e}_t) \leq -\varepsilon_3 e^T(t)e(t) < 0, t \neq kh + \tau_k, k \in \mathbb{N}$ for $e(t) \neq 0$, which implies that (2.10) is satisfied. By utilizing Lemma 2.2, one can conclude that the system (2.6)–(2.7) is globally asymptotically stable.

Note that Σ is a convex combination of $\Upsilon_{11} + \Omega_2$ and Ω_3 on $\tau(t) \in [\alpha, \beta]$. Therefore, $\Sigma < 0$ if the following inequalities hold

$$\Pi_{11} + \Omega_1 + \sigma(\Upsilon_{11} + \Omega_2) < 0 \tag{2.21}$$

$$\Pi_{11} + \Omega_1 + \sigma\Omega_3 < 0 \tag{2.22}$$

Applying the Schur complement to (2.21) and (2.22), we arrive at (2.15) and (2.16), respectively. The proof is completed. □

It is clear that Theorem 2.1 establishes the relationship among the network topology, the consensus controller gain and both upper and lower bounds of network-induced delays. Since $\beta = h + \max_k\{\tau_k \mid k \in \mathbb{N}\}$, it is clear that $\tau_k \leq \beta - h$. When β is known, there is a tradeoff between the sampling period and the allowable network-induced delay, i.e., a smaller sampling period for the distributed multi-agent system can tolerate a larger network-induced delay. It is worth pointing out that in [6], a sampled-data consensus protocol was proposed, however, the communication delays are required to be smaller than the sampling period h, which is in fact involved in the consensus protocol (2.4) as a special case. Hence, to some extent, the consensus protocol (2.4) can be regarded as a more general one. Furthermore, assuming that $\tau_k = 0$ $(k \in \mathbb{N})$, the system (2.6) can be reduced to a sampled-data multi-agent system, which is similar to [7]. In such a case, one can derive the corresponding result by setting $\alpha = 0$ in Theorem 2.1. Then, by employing this result, one can obtain the maximum allowable sampling period.

2.4 Network-Based Consensus Controller Design

Based on Theorem 2.1, we are in a position to give a sufficient condition on the existence of the networked-based consensus controller for the distributed multi-agent system.

Theorem 2.2 *Under Assumptions 2.2.1 and 2.2.2, given some scalars* $\varepsilon > 0$, $\mu > 0$, α *and* β *with* $\beta \geq \alpha$, *the leader-following consensus for the system described by* (2.1) *and* (2.2) *can be asymptotically achieved by the control law* (2.4) *if there exist real matrices* $\hat{P} > 0$, $\hat{Q}_1 > 0$, $\hat{Q}_2 > 0$, $\hat{R}_1 > 0$, $\hat{R}_2 > 0$, $\hat{Y} > 0$, *and matrices* $\hat{Z}_k$ $(k = 1, 2, 3)$ *of appropriate dimensions such that*

$$\begin{bmatrix} \hat{\Pi}_{11} + \sigma\hat{\Upsilon}_{11} & \hat{\Pi}_{12} & \sigma\hat{Z}_1^T \\ * & \hat{\Pi}_{22} + \hat{\Upsilon}_{22} & 0 \\ * & * & -\sigma U \otimes \hat{R}_2 \end{bmatrix} < 0 \tag{2.23}$$

$$\begin{bmatrix} \hat{\Pi}_1 & \hat{\Pi}_{12} & \sigma\hat{Z}_2^T & \sigma\hat{Z}_3^T \\ * & \hat{\Pi}_{22} & 0 & 0 \\ * & * & -\sigma U \otimes \hat{R}_2 & 0 \\ * & * & * & -\sigma U \otimes \hat{Y} \end{bmatrix} < 0 \tag{2.24}$$

where

$$
\begin{aligned}
\hat{\Pi}_{11} &= \mathscr{E}_1^T[(U \otimes A\hat{P})\mathscr{E}_1 - (UH \otimes B\hat{K})\mathscr{E}_2] \\
&\quad + [(U \otimes A\hat{P})\mathscr{E}_1 - (UH \otimes B\hat{K})\mathscr{E}_2]^T \mathscr{E}_1 \\
&\quad + \mathscr{E}_1^T (U \otimes \hat{Q}_1)\mathscr{E}_1 + \mathscr{E}_3^T (U \otimes (\hat{Q}_2 - \hat{Q}_1))\mathscr{E}_3 \\
&\quad - \mathscr{E}_4^T (U \otimes \hat{Q}_2)\mathscr{E}_4 - \mathscr{E}_{13}^T (U \otimes \hat{R}_1)\mathscr{E}_{13} \\
&\quad - \mathscr{E}_{15}^T(\varepsilon U \otimes \hat{P})\mathscr{E}_{15} - \mathscr{E}_{24}^T \hat{Z}_1 - \hat{Z}_1^T \mathscr{E}_{24} \\
&\quad - \mathscr{E}_{32}^T \hat{Z}_2 - \hat{Z}_2^T \mathscr{E}_{32} - \mathscr{E}_{15}^T \hat{Z}_3 - \hat{Z}_3^T \mathscr{E}_{15}, \\
\hat{\Pi}_{12} &= [(U \otimes A\hat{P})\mathscr{E}_1 - (UH \otimes B\hat{K})\mathscr{E}_2]^T, \\
\hat{\Pi}_{22} &= U \otimes [\mu^2(\alpha^2 \hat{R}_1 + \sigma \hat{R}_2) - 2\mu \hat{P}], \\
\hat{\Upsilon}_{11} &= \mathscr{E}_{15}^T[(\varepsilon U \otimes A\hat{P})\mathscr{E}_1 - (\varepsilon UH \otimes B\hat{K})\mathscr{E}_2] \\
&\quad + [(\varepsilon U \otimes A\hat{P})\mathscr{E}_1 - (\varepsilon UH \otimes B\hat{K})\mathscr{E}_2]^T \mathscr{E}_{15}, \\
\hat{\Upsilon}_{22} &= \sigma \mu^2 U \otimes \hat{Y}.
\end{aligned}
$$

Moreover, the controller gain is given by

$$
K = \hat{K}\hat{P}^{-1}. \tag{2.25}
$$

Proof Define $\hat{P} = P^{-1}$, $\hat{Q}_1 = P^{-1}Q_1P^{-1}$, $\hat{Q}_2 = P^{-1}Q_2P^{-1}$, $\hat{R}_1 = P^{-1}R_1P^{-1}$, $\hat{R}_2 = P^{-1}R_2P^{-1}$, $\hat{Y} = P^{-1}YP^{-1}$, $X = \varepsilon P$, $J = diag\{I_N \otimes P^{-1}, \ldots, I_N \otimes P^{-1}\}$, $\hat{Z}_k = JZ_kJ(k = 1, 2, 3)$ and $\hat{K} = KP^{-1}$. Pre- and post-multiplying both the sides of (2.15) by $diag\{J, I_N \otimes (\alpha^2 R_1 + \sigma R_2 + \sigma Y)^{-1}, I_N \otimes P^{-1}\}$, and both the sides of (2.16) by $diag\{J, I_N \otimes (\alpha^2 R_1 + \sigma R_2)^{-1}, I_N \otimes P^{-1}, I_N \otimes P^{-1}\}$, respectively, one can obtain (2.23) and (2.24), where the term $U \otimes [P^{-1}(\alpha^2 \hat{R}_1 + \sigma \hat{R}_2 + \sigma \hat{Y})^{-1} P^{-1}]$ is resolved by the inequality $-\hat{P}\hat{R}^{-1}\hat{P} \leq \mu^2 \hat{R} - 2\mu \hat{P}$. The proof of Theorem 2.2 is completed. □

Remark 2.3 Note that, in [2] the consensus controller is required to be given **a priori**. However, in this chapter, such the requirement is no longer needed. Employing Theorem 2.2, one can design an appropriate consensus controller.

Remark 2.4 In the case of $N = 1$, one can view the leader agent and the follower agent as a "master" system and a "slave" system, respectively. In this case, the corresponding master-slave synchronization result can be derived.

2.5 A Numerical Example

Consider the system (2.1) with double-integrator agents described by

$$
\dot{x}_i(t) = \begin{bmatrix} 0 & 1 \\ 0 & 0 \end{bmatrix} x_i(t) + \begin{bmatrix} 0 \\ 1 \end{bmatrix} u_i(t) \quad (i = 1, 2, 3) \tag{2.26}
$$

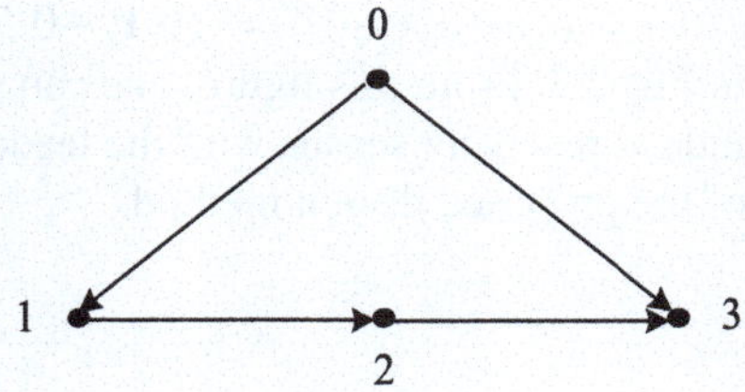

Fig. 2.2 The graph $\mathcal{G}$ for the multi-agent system

Table 2.1 MAUB of $\tau(t)$ for different α

α	0.01	0.05	0.10	0.20	0.30
β	0.440	0.442	0.444	0.445	0.445

under a directed graph in Fig. 2.2. For simplicity, it is assumed that all the weights are set as 1.

First, we analyze the stability. Let the controller gain matrix $K = [0.1, 1]$. Employing Theorem 2.1, we calculate the maximum allowable upper bound (MAUB) β of $\tau(t)$ for various lower bound α and list them in Table 2.1. As mentioned, there is a tradeoff between the sampling period and the allowable network-induced delay. For example, let $\alpha = 0.05$ and $h = 0.1$. Then the network-induced delay τ_k satisfies $0.05 \leq \tau_k \leq 0.342$ for $k \in \mathbb{N}$.

Second, we design the network-based controller. Set the sample period $h = 0.1$ and let the lower bound and upper bound of network-induced delay τ_k be 0.1 and 0.2, respectively. Then it is easy to obtain $\alpha = 0.1$ and $\beta = 0.3$. By solving linear matrix inequalities (LMIs) in Theorem 2.2 with $\varepsilon = 2$ and $\mu = 1$, we obtain the controller gain as $K = \begin{bmatrix} 0.07 & 0.6042 \end{bmatrix}$. Choosing the initial states of the corresponding error system as $e_1(0) = (e_{11}(0), e_{12}(0)) = (0.5, 1)$, $e_2(0) = (e_{21}(0), e_{22}(0)) = (0, 1)$ and

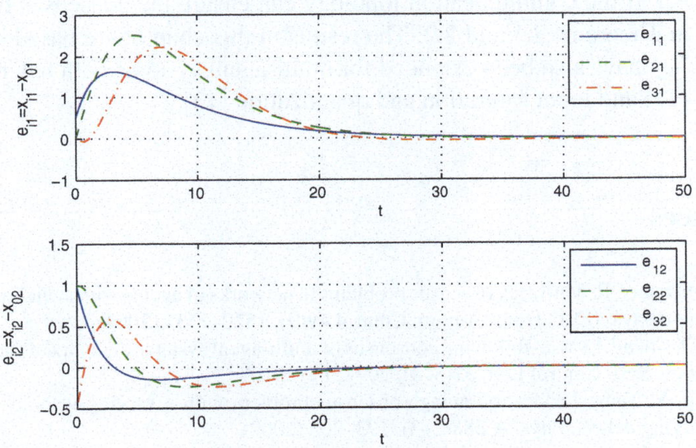

Fig. 2.3 The response of the error system

$e_3(0) = (e_{31}(0), e_{32}(0)) = (0.1, -0.5)$, we depict the response of the error system in Fig. 2.3. From this figure, one can see clearly that the follower agents' states can indeed reach consensus with the leader agent's state, which shows the effectiveness of the proposed design method.

2.6 Concluding Remarks

This chapter has studied network-based leader-following consensus and proposed a network-based consensus control protocol for distributed multi-agent systems. A Lyapunov-Krasovskii functional with digraph information has been constructed to derive a delay-dependent sufficient condition, which makes the follower agents' states reach an agreement on the leader's state. Based on this condition, the corresponding controller gain has been obtained by solving a set of LMIs. An example has been given to illustrate the effectiveness of the derived results.

2.7 Notes

Notice that, in the literature [7–9], an eigenvalue-based method was used to conduct the consensus analysis. However, due to the presence of time-varying communication delays, such an eigenvalue-based method is no longer applicable to deal with the issue of network-based consensus considered in this chapter. Instead, a Lyapunov-Krasovskii functional approach like [4, 5, 10] is applied to obtain sufficient conditions for achieving the analysis of network-based leader-following consensus and the design of network-based controllers, respectively. Since the Laplacian matrix of communication graph is included in the conditions of Theorems 2.1 and 2.2, in fact we can check if the communication topology can ensure the consensus by solving the LMIs in Theorems 2.1 and 2.2. The results in this chapter are based mainly on [11], and they have also been extended for multi-agent systems with heterogeneous dynamics [12] and input saturation and quantization [13].

References

1. R. Olfati-Saber, R. Murray, Consensus problems in networks of agents with switching topology and time-delays. IEEE Trans. Autom. Control **49**(9), 1520–1533 (2004)
2. W. Ni, D. Cheng, Leader-following consensus of multi-agent systems under fixed and switching topologies. Syst. Control Lett. **57**(3–4), 209–217 (2010)
3. K. Peng, Y. Yang, Leader-following consensus problem with a varying-velocity leader and time-varying delays. Phys. A **388**(2–3), 193–208 (2009)
4. Q.-L. Han, Absolute stability of time-delay systems with sector-bounded nonlinearity. Automatica **41**(12), 2171–2176 (2005)

5. X.-M. Zhang, Q.-L. Han, Novel delay-derivative-dependent stability criteria using new bounding techniques. Int. J. Robust Nonlinear Control **23**(13), 1419–1432 (2013)
6. Y. Zhang, Y.P. Tian, Consensus of data-sampled multi-agent systems with random communication delay and packet loss. IEEE Trans. Autom. Control **55**(4), 939–943 (2010)
7. W. Yu, W.X. Zheng, G. Chen, W. Ren, J. Cao, Second-order consensus in multi-agent dynamical systems with sampled position data. Automatica **47**(7), 1496–1503 (2011)
8. W. Yu, G. Chen, W. Ren, J. Kurths, W. X. Zheng, Distributed higher order consensus protocols in multiagent dynamical systems. IEEE Trans. Circuits Syst. I: Regul. Pap. **58**(8), 1924–1932 (2011)
9. K. You, L. Xie, Network topology and communication data rate for consensusability of discrete-time multi-agent systems. IEEE Trans. Autom. Control **56**(10), 2262–2275 (2011)
10. E. Fridman, A refined input delay approach to sampled-data control. Automatica **46**(2), 421–427 (2010)
11. L. Ding, Q.-L. Han, G. Guo, Network-based leader-following consensus for distributed multi-agent systems. Automatica **49**(7), 2281–2286 (2013)
12. L. Ding, W.X. Zheng, Network-based practical consensus of heterogeneous nonlinear multiagent systems. IEEE Trans. Cybern. **47**(8), 1841–1851 (2017)
13. L. Ding, W.X. Zheng, G. Guo, Network-based practical set consensus for multi-agent systems subject to input saturation. Automatica **89**(3), 316–324 (2018)

Chapter 3
Sampled-Data-Based Event-Triggered Consensus of Multi-agent Systems

This chapter is concerned with event-triggered sampled-data consensus for distributed multi-agent systems with a directed graph. A distributed event-triggered sampled-data transmission strategy is proposed, which allows the event-triggering condition to be intermittently examined at constant sampling instants. Based on this strategy, a sampled-data consensus control protocol is presented, with which the consensus of distributed multi-agent systems can be transformed into the stability of a system with a time-varying delay. Then, a sufficient condition on consensus of the multi-agent system is derived. Correspondingly, a co-design algorithm for obtaining both the parameters of the distributed event-triggered transmission strategy and the consensus controller gain is proposed. Two numerical examples are given to show the effectiveness of the proposed method.

3.1 Sampled-Data-Based Event-Triggered Communication Schemes

Consider a distributed multi-agent system consisting of N agents, where each agent's dynamics can be described by

$$\dot{x}_i(t) = Ax_i(t) + Bu_i(t), \quad i = 1, 2, \ldots, N. \tag{3.1}$$

where $A \in \mathbb{R}^{n\times n}$ and $B \in \mathbb{R}^{n\times m}$ are constant matrices; $x_i \in \mathbb{R}^n$ and $u_i \in \mathbb{R}^m$ are the state and the control input of agent i, respectively. The initial condition is given by $x_i(0) = x_i^0$. It is assumed that the agents can communicate with their neighbors via a communication graph $\mathscr{G}$.

L. Ding et al., *Distributed Control and Optimization of Networked Microgrids*,
Power Systems, https://doi.org/10.1007/978-3-030-95029-3_3

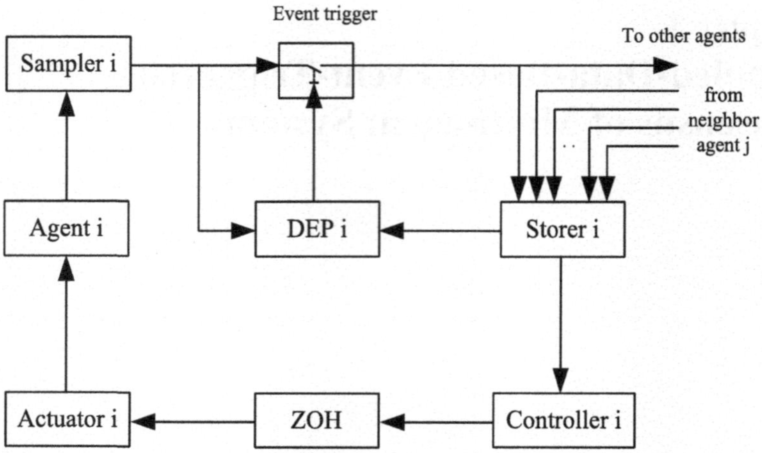

Fig. 3.1 A distributed event-triggered transmission strategy for agent i

The purpose of this chapter is to design an event-triggered transmission strategy such that the system (3.1) can asymptotically achieve consensus, i.e.,

$$\lim_{t\to\infty} \|x_i(t) - x_j(t)\| = 0$$

for any initial condition $x_i(0)$ and $i, j = 1, 2, \ldots, N$.

For this purpose, we present the distributed event-triggered transmission strategy for the distributed multi-agent system, whose framework is illustrated in Fig. 3.1. In this framework, whether or not sampled-data of agent i should be broadcasted at the sampling instant kh ($k \in \mathbb{N}$) depends on whenever its threshold is violated. More specifically, it is assumed that the sampler is time-driven, the storer, the controller and the actuator with ZOH are all event-driven and all the sampled-data packets are transmitted with their time stamps. The state of each agent i is sampled by a sampler at a constant sampling period $h > 0$. The sampled-data $x_i(kh)$ ($k \in \mathbb{N}$) of agent i can be successfully sent to its distributed event processor (DEP) embedding an event-triggered condition. Here, the DEP's execution of each agent i involves three tasks:

(i) Receiving the sampled-data from agent i and its storer at each sampling instant;
(ii) Processing the received sampled-data with respect to the event-triggered condition; and
(iii) Generating a trigger signal to the event trigger when a threshold is violated.

Once the event trigger receives a trigger signal, the sampled-data of agent i is authorized to be broadcasted to the storers of its own and its neighbors'. It should be pointed out that the storer is employed to gather the sampled-data of agent i and its neighbors and to immediately update its information as soon as measuring the sampled-data from agent i and its neighbors. Then, the controller of agent i updates its input by using the received sampled-data from the storer, which implies that the controller update of agent i is driven at both its own and neighbors' event instants. Then, the output of the controller of agent i is sent to the actuator with ZOH, where ZOH is used to keep the control input of agent i constant when no newest sampled-data from the controller of agent i arrives.

Note that, whether or not the sampled-data of agent i should be transmitted between neighbor agents relies on an event-triggered condition rather than the elapse of a fixed time. Denote the mth transmitted instant of the sampled-data of agent i by $t_m^i h$, $m \in \mathbb{N}$. Hence, the next broadcasting instant $t_{m+1}^i h$ of agent i is accordingly determined by

$$\begin{aligned} t_{m+1}^i h = t_m^i h + & \min_{l_i \geq 1} \{l_i h | \chi_i^T (t_m^i h + l_i h) \Phi \chi_i (t_m^i h + l_i h) \\ & \geq \sigma_i y_i^T (t_m^i h + l_i h) \Phi y_i (t_m^i h + l_i h)\} \end{aligned} \tag{3.2}$$

where $l_i \in \mathbb{N}$, $\sigma_i > 0$ is a threshold parameter, $\Phi > 0$ is a weighting matrix,

$$\chi_i (t_m^i h + l_i h) = x_i (t_m^i h + l_i h) - x_i (t_m^i h)$$

and

$$y_i (t_m^i h + l_i h) = \sum_{j=1}^{N} a_{ij} [x_i (t_m^i h) - x_j (t_{m_j'}^j h)]$$

with $m_j' \triangleq \arg\min_p \{t_m^i + l_i - t_p^j | t_m^i + l_i > t_p^j,\ p \in \mathbb{N}\}$ with $\arg\min_p f(q)$ is the value of q for which $f(q)$ attains its minimum value.

Apparently, one can see from (3.2) that, at the kth sampling instant, the event-triggered condition for agent i is closely related to the sampled-data error $\chi_i (t_m^i h + l_i h)$ and the sampled-data $y_i (t_m^i h + l_i h)$ including the latest transmitted sampled-data $x_i (t_m^i h)$ of agent i and the latest transmitted sampled-data $x_j (t_{m_j'}^j h)$ of its neighbors. Then, we propose the following sampled-data consensus protocol

$$u_i (t) = -K \sum_{j=1}^{N} a_{ij} [x_i (t_m^i h) - x_j (t_{m_j'(t)}^j h)], \tag{3.3}$$

where $t \in [t_m^i h, t_{m+1}^i h)$, $m_j'(t) \triangleq \arg\min_p \{t - t_p^j | t \geq t_p^j,\ p \in \mathbb{N}\}$ and K is the consensus controller gain to be designed later.

Remark 3.1 It can be intuitionally seen from (3.2) that the the next broadcasting instant $t_{m+1}^i h$ of agent i is not only related to its own sampled-data but also to its neighbors' sampled-data. If $\chi_i^T (t_m^i h + l_i h) \Phi \chi_i (t_m^i h + l_i h) < \sigma_i y_i^T (t_m^i h + l_i h) \Phi y_i (t_m^i h + l_i h)$, which means that the sampled-data error still belongs to an allowable range, then the sampled-data of agent i is unnecessarily transmitted to its neighbor agents. In this sense, the frequency of sample-data transmission and control input update may be significantly reduced by the event-triggered transmission strategy (3.2), which implies that the communication and computation resources can be saved.

Remark 3.2 It is worth mentioning that when the threshold $\sigma_i \to 0^+$ for $i = 1, 2, \ldots, N$, one can obtain that $\chi_i^T(t_m^i h + l_i h)\Phi\chi_i(t_m^i h + l_i h) \geq 0$ for $i = 1, 2, \ldots, N$, which means that the sampled-data of agent i is broadcasted at each sampling time. In such a case, the distributed event-triggered transmission strategy is reduced to the time-triggered transmission strategy [1–3].

Remark 3.3 It should be pointed out that compared with the continuous event-triggered transmission strategy in [4], the "discrete" event-triggered transmission strategy proposed in this chapter is mainly involved with two aspects: (i) for agent i, only the broadcasted neighbors' sampled-data $x_j(t_m^j h)$ is used rather than the continuous states $x_j(t)$; and (ii) the event-triggered condition is only required to be examined at each sampling instant, which implies that an extra hardware for the continuous measurement and computation is no longer needed.

Remark 3.4 In [4] and [5], a lower bound of inter-event times was specially analyzed in order to guarantee that no Zeno-behavior occurs. In this chapter, however, it is easily seen that the inter-event time is at least lower bounded by the constant sampling period $h > 0$, which implies that Zeno-behavior is absolutely excluded.

3.2 Consensus Analysis and Synthesis

3.2.1 Model Transformation

Define the error between agent i's sampled-data at the kth sampling time and its last triggered time $t_m^i h$ as

$$e_i(kh) \triangleq x_i(kh) - x_i(t_m^i h), \quad t_m^i \leq k < t_{m+1}^i \tag{3.4}$$

Remark 3.5 From the definition of $\chi_i(kh)$ and $e_i(kh)$, one can clearly see that, when $k \neq t_m^i$, $\chi_i(kh) = e_i(kh)$, however, for $k = t_m^i$, $\chi_i(t_m^i h) = x_i(t_m^i h) - x_i(t_{m-1}^i h)$ and $e_i(t_m^i h) = 0$.

Note that the release time interval $[t_m^i h, t_{m+1}^i h) = \bigcup_{k=t_m^i}^{t_{m+1}^i - 1} [kh, (k+1)h)$. Hence, by dividing $[t_m^i h, t_{m+1}^i h)$ into $t_{m+1}^i - t_m^i$ sampling interval and from (3.1), (3.3) and (3.4), we have the closed-loop multi-agent system for $t \in [kh, (k+1)h)$

$$
\begin{aligned}
\dot{x}_i(t) = &\ Ax_i(t) - BK\sum_{j=1}^{N} a_{ij}[x_i(t_m^i h) - x_j(t^j_{m'_j(t)} h)] \\
= &\ Ax_i(t) - BK\sum_{j=1}^{N} a_{ij}[x_i(kh) - x_j(kh) \\
&\ - e_i(kh) + e_j(kh)],
\end{aligned} \tag{3.5}
$$

Define the error variables $z_i(t) = x_1(t) - x_i(t)$ and $\mathscr{E}_i(kh) = e_1(kh) - e_i(kh)$ for $i = 2, \ldots, N$, and the vector

$$
\begin{aligned}
x(t) &= [x_1^T(t), x_2^T(t), \ldots, x_N^T(t)]^T, \\
e(kh) &= [e_1^T(kh), e_2^T(kh), \ldots, e_N^T(kh)]^T, \\
z(t) &= [z_2^T(t), z_3^T(t), \ldots, z_N^T(t)]^T, \\
\mathscr{E}(kh) &= [\mathscr{E}_2^T(kh), \mathscr{E}_3^T(kh), \ldots, \mathscr{E}_N^T(kh)]^T.
\end{aligned}
$$

According to [6], with $E_1 = [\mathbf{1}, -I_{N-1}]$ and $E_2 = [\mathbf{0}, -I_{N-1}]^T$, one obtains

$$
\begin{aligned}
z(t) = (E_1 \otimes I_n)x(t),&\ \ x(t) = (E_2 \otimes I_n)z(t) + (\mathbf{1} \otimes I_n)x_1(t) \\
\mathscr{E}(kh) = (E_1 \otimes I_n)e(kh),&\ \ e(kh) = (E_2 \otimes I_n)\mathscr{E}(kh) + (\mathbf{1} \otimes I_n)e_1(kh).
\end{aligned}
$$

Then, we have, for $t \in [kh, (k+1)h)$

$$
\dot{z}(t) = (I_{N-1} \otimes A)z(t) - (\mathscr{L} \otimes BK)z(kh) + (\mathscr{L} \otimes BK)\mathscr{E}(kh) \tag{3.6}
$$

where $\mathscr{L} = E_1 L E_2 \in \mathbb{R}^{(N-1)\times(N-1)}$.

Define an "artificial delay" $\tau(t) = t - kh, t \in [kh, (k+1)h)$. Clearly, $\tau(t)$ is piecewise-linear with the derivative $\dot{\tau}(t) = 1$ at $t \neq kh$ and is discontinuous at $t = kh$. It is easily known that $0 \leq \tau(t) < h$. Hence, the system (3.6) can be written as

$$
\begin{aligned}
\dot{z}(t) = &\ (I_{N-1} \otimes A)z(t) - (\mathscr{L} \otimes BK)z(t - \tau(t)) \\
&\ + (\mathscr{L} \otimes BK)\mathscr{E}(t - \tau(t)), \quad t \in [kh, (k+1)h)
\end{aligned} \tag{3.7}
$$

The initial condition of the state $z(t)$ is supplemented as $z(\theta) = \phi(\theta), \theta \in [-h, 0]$ with $\phi(0) = z(0) = [z_2^T(0), z_3^T(0), \ldots, z_N^T(0)]^T$ and $\phi \in W$, where W denotes the Banach space of absolutely continuous functions $[-h, 0] \to \mathbb{R}^{(N-1)n}$ with square-integrable derivatives and the norm

$$
\|\phi\|_W = \max_{\theta \in [-h,0]} \|\phi(\theta)\| + \left[\int_{-h}^{0} \|\dot{\phi}(s)\|^2 ds\right]^{\frac{1}{2}}.
$$

Similar to the proof in [6], we have the following result.

Theorem 3.1 *Suppose that the directed graph $\mathscr{G}$ has a spanning tree. Then, under the event-triggered transmission strategy (3.2), consensus of the multi-agent system (3.1) with (3.3) can be asymptotically achieved if and only if the system (3.7) is asymptotically stable.*

From Theorem 3.1, one can clearly see that consensus of multi-agent systems (3.1) can be converted into the stability problem of (3.7). Next, we analyze the stability of (3.7). Before ending this subsection, a useful lemma is given for the proofs of main results.

Lemma 3.1 *Under the event-triggered transmission strategy (3.2), the following inequality holds for any $k \in \mathbb{N}$*

$$e_i^T(kh)\Phi e_i(kh) \le \sigma_i \tilde{y}_i^T(kh)\Phi \tilde{y}_i(kh) \tag{3.8}$$

where $\tilde{y}_i(kh) = \sum_{j=1}^{N} a_{ij}[x_i(t^i_{m_i(k)}h) - x_j(t^j_{m_j(k)}h)]$ *with* $m_i(k) \triangleq \arg\min_p\{k - t^i_p | k \ge t^i_p,\ p \in \mathbb{N}\}$.

Proof It is assumed that $t^i_m \le k < t^i_{m+1}$. For $k = t^i_m$, it is true that $e_i(kh) = 0$, and thus $\sigma_i \tilde{y}_i^T(kh)\Phi \tilde{y}_i(kh) \ge 0$. For $t^i_m < k < t^i_{m+1}$, it is easy to know that $e_i(kh) = \chi_i(kh)$, $\tilde{y}_i(kh) = y_i(kh)$ and there is no event occurring at the kth sampling times. According to the property of the event-triggered strategy (3.2), we have that $e_i^T(kh)\Phi e_i(kh) < \sigma_i \tilde{y}_i^T(kh)\Phi \tilde{y}_i(kh)$. Thus, it can be concluded that at any sampling time kh the inequality (3.8) holds. The proof is completed. □

3.2.2 Stability Analysis and Controller Design

In this subsection, we first derive a sufficient condition on the stability of the system (3.7) under the event-trigged transmission strategy (3.2). Then, based on this condition, we give a sufficient condition on the existence of consensus controller gain.

Choose a Lyapunov-Krasovskii functional as

$$V(t, z_t, \dot{z}_t) = \sum_{i=1}^{2} V_i(t, z_t, \dot{z}_t),\ \forall t \in [kh, (k+1)h) \tag{3.9}$$

where

$$V_1(t, z_t, \dot{z}_t) = z^T(t)Pz(t) + \int_{t-h}^{t} z^T(s)Qz(s)ds + \int_{-h}^{0}\int_{t+\theta}^{t} \dot{z}^T(s)R\dot{z}(s)dsd\theta$$

$$V_2(t, z_t, \dot{z}_t) = (h - \tau(t))\{[z(t) - z(t - \tau(t))]^T X[z(t) - z(t - \tau(t))] + \int_{t-\tau(t)}^{t} \dot{z}^T(s)Y\dot{z}(s)ds\}$$

with $z_t(\theta) = z(t+\theta), \theta \in [-h, 0]$, $P > 0$, $Q > 0$, $R > 0$, $X > 0$ and $Y > 0$. For the sake of simplicity, let $\psi(t) = [z^T(t), z^T(t - \tau(t)), z^T(t - h), \mathscr{E}^T(t - \tau(t))]^T$ and $\mathscr{E}_i$ be a block entry matrix with $\mathscr{E}_{ij} = \mathscr{E}_i - \mathscr{E}_j$. For example, $\mathscr{E}_2 = [0, I_{(N-1)n}, 0, 0]$ and $\mathscr{E}_{24} = [0, I_{(N-1)n}, 0, -I_{(N-1)n}]$.

In order to proceed, we need the following lemmas.

Lemma 3.2 ([7]) *Let there exist positive numbers $\varepsilon_i (i = 1, 2, 3)$ and a Lyapunov functional $V(t, z_t, \dot{z}_t) : \mathbb{R} \times \mathbb{R}^{Nn} \times \mathbb{R}^{Nn} \to \mathbb{R}$ such that $V(t, z_t, \dot{z}_t)$ is continuous from the right for $z(t)$ satisfying (3.7), absolutely continuous and differentiable for $t \neq kh$ and satisfies*

$$\varepsilon_1 \|z(t)\|^2 \leq V(t, z_t, \dot{z}_t) \leq \varepsilon_2 \|z_t\|^2 \tag{3.10}$$

$$\lim_{t \to kh^-} V(t, z_t, \dot{z}_t) \geq V(t, z_t, \dot{z}_t)|_{t=kh} \tag{3.11}$$

$$\dot{V}(t, z_t, \dot{z}_t) \leq \varepsilon_3 \|z(t)\|^2, \quad t \neq kh. \tag{3.12}$$

Then, system (3.7) is asymptotically stable.

Now, we establish the following result.

Theorem 3.2 *Given $h > 0$ and $\Lambda = diag\{\sigma_1, \sigma_2, \ldots, \sigma_N\}$, the system (3.7) under the event-triggered transmission strategy (3.2) is globally asymptotically stable if the graph $\mathscr{G}$ has a directed spanning tree and there exist real matrices $P > 0$, $Q > 0$, $R > 0$, $X > 0$, $Y > 0$, and some matrices M_k $(k = 1, 2)$ of appropriate dimensions such that*

$$\begin{bmatrix} \Gamma_{11} + h\Xi_{11} & hF^T(R+Y) & hM_1^T \\ * & -h(R+Y) & 0 \\ * & * & -hR \end{bmatrix} < 0 \tag{3.13}$$

$$\begin{bmatrix} \Gamma_{11} & hF^TR & hM_2^T \\ * & -hR & 0 \\ * & * & -h(R+Y) \end{bmatrix} < 0 \tag{3.14}$$

where

$$\begin{aligned}
\Gamma_{11} = {} & \mathscr{E}_1^T PF + FP\mathscr{E}_1 - \mathscr{E}_{12}^T X \mathscr{E}_{12} + \mathscr{E}_1^T Q \mathscr{E}_1 - \mathscr{E}_3^T Q \mathscr{E}_3 \\
& -\mathscr{E}_{23}^T M_1 - M_1^T \mathscr{E}_{23} - \mathscr{E}_{12}^T M_2 - M_2^T \mathscr{E}_{12} \\
& +\gamma \mathscr{E}_{24}^T (E_2^T L^T \Lambda L E_2 \otimes \Phi)\mathscr{E}_{24} - \mathscr{E}_4^T (I_{N-1} \otimes \Phi)\mathscr{E}_4, \\
\Xi_{11} = {} & \mathscr{E}_{12}^T X F + F^T X \mathscr{E}_{12}, \\
F = {} & (I_{N-1} \otimes A)\mathscr{E}_1 - (\mathscr{L} \otimes BK)\mathscr{E}_{24},
\end{aligned}$$

with γ *is the largest eigenvalue of* $E_1^T E_1$.

Proof It is clear that $V(t, z_t, \dot{z}_t)$ satisfies (3.10). Since $V_2(t, z_t, \dot{z}_t) \geq 0$ and $V_2(t, z_t, \dot{z}_t) = 0$ at $t = kh$, one can see that the condition (3.11) holds. Taking the derivative of (3.9) along the trajectory (3.7) with respect to $t \in (kh, (k+1)h)$ and using the fact that $\dot{\tau}(t) = 1$ for $t \neq kh$, we have

$$\begin{aligned}
\dot{V}(t, z_t, \dot{z}_t) = {} & \psi^T(t)\{2\mathscr{E}_1^T PF + \mathscr{E}_1^T Q \mathscr{E}_1 - \mathscr{E}_3^T Q \mathscr{E}_3 + hF^T RF - \mathscr{E}_{12}^T X \mathscr{E}_{12} \\
& + (h - \tau(t))[2\mathscr{E}_{12}^T X + FY]^T F - \mathscr{E}_4^T (I_{N-1} \otimes \Phi)\mathscr{E}_4\}\psi^T(t) \\
& + \eta_1 + \eta_2 + \mathscr{E}(kh)^T (I_{N-1} \otimes \Phi)\mathscr{E}(kh)
\end{aligned}$$

where

$$\begin{aligned}
\eta_1 &= -\int_{t-h}^{t} \dot{z}^T(s) R \dot{z}(s) ds, \\
\eta_2 &= -\int_{t-\tau(t)}^{t} \dot{z}^T(s) Y \dot{z}(s) ds.
\end{aligned}$$

Note that

$$\begin{aligned}
\eta_1 + \eta_2 = {} & -\int_{t-h}^{t-\tau(t)} \dot{z}^T(s) R \dot{z}(s) ds \\
& -\int_{t-\tau(t)}^{t} \dot{z}^T(s)(R + Y)\dot{z}(s) ds
\end{aligned} \tag{3.15}$$

and $\int_{t-h}^{t-\tau(t)} \dot{z}(s) ds = \mathscr{E}_{23}\psi(t)$ and $\int_{t-\tau(t)}^{t} \dot{z}(s) ds = \mathscr{E}_{12}\psi(t)$.

Applying Lemma 2.4, we obtain

$$\begin{aligned}
\eta_1 + \eta_2 \leq {} & \psi^T(t)\{(h - \tau(t))M_1^T R^{-1} M_1 - \mathscr{E}_{23}^T M_1 \\
& - M_1^T \mathscr{E}_{23} + \tau(t) M_2^T (R + Y)^{-1} M_2 \\
& - \mathscr{E}_{12}^T M_2 - M_2^T \mathscr{E}_{12}\}\psi(t)
\end{aligned} \tag{3.16}$$

Due to the fact that

$$\begin{aligned}
\mathscr{E}(kh)^T (I_{N-1} \otimes \Phi)\mathscr{E}(kh) &= e^T(kh)(E_1^T E_1 \otimes \Phi)e(kh) \\
&\leq \gamma e^T(kh)(I_N \otimes \Phi)e(kh)
\end{aligned}$$

and applying Lemma 3.1, one obtains

$$\begin{aligned}&\mathscr{E}(kh)^T(I_{N-1}\otimes\Phi)\mathscr{E}(kh)\\ \le\ &\gamma[z(kh)-\mathscr{E}(kh)]^T(E_2^TL^T\Lambda LE_2\otimes\Phi)[z(kh)-\mathscr{E}(kh)]\end{aligned} \tag{3.17}$$

From (3.15) to (3.17), we have

$$\dot{V}(t,e_t,\dot{e}_t)\le\psi^T(t)\Sigma\psi(t),\quad t\in(kh,(k+1)h)$$

where

$$\begin{aligned}&\Sigma=\Gamma_{11}+\Omega_1+(h-\tau(t))(\Xi_{11}+\Omega_2)+\tau(t)\Omega_3,\\ &\Omega_1=hF^TRF,\ \Omega_2=F^TYF+M_1^TR^{-1}M_1,\\ &\Omega_3=M_2^T(R+Y)^{-1}M_2\end{aligned}$$

It is clear that if $\Sigma<0$, then there exists a scalar $\varepsilon_3>0$ such that $\dot{V}(t,e_t,\dot{e}_t)\le -\varepsilon_3 z^T(t)z(t), t\ne kh, k\in\mathbb{N}$ for $z(t)\ne 0$. By using Lemma 3.2, one can conclude that the system (3.7) is globally asymptotically stable. Notice that Σ is a convex combination of $\Xi_{11}+\Omega_2$ and Ω_3 on $\tau(t)\in[0,h]$. Therefore, $\Sigma<0$ if

$$\Gamma_{11}+\Omega_1+h(\Xi_{11}+\Omega_2)<0 \tag{3.18}$$

$$\Gamma_{11}+\Omega_1+h\Omega_3<0 \tag{3.19}$$

Applying the Schur complement to (3.18) and (3.19), we arrive at (3.13) and (3.14), respectively. The proof is completed. □

Remark 3.6 Note that different from [4], the network topology of the multi-agent system is extended to a more general directed graph with a spanning tree rather than an undirected connected graph.

Remark 3.7 Theorem 3.2 establishes the relationship among network topologies, the event-triggered transmission strategy, the sampling period and the consensus controller gain. For example, given the network topology L, the parameters σ_i of the event-triggered transmission strategy and the control gain K, one can obtain the allowable maximum of sampling period h by solving a set of linear matrix inequalities (LMIs) in (3.13) and (3.14).

Based on Theorem 3.2, a sufficient condition on the existence of a consensus controller is given below for the distributed multi-agent system under the event-triggered transmission strategy (3.2).

Theorem 3.3 *Given scalars $h>0$, $\mu>0$, $\delta>0$ and $\Lambda=diag\{\sigma_1,\sigma_2,\ldots,\sigma_N\}$, consensus of the distributed multi-agent system (3.1) with (3.3) under the event-triggered transmission strategy (3.2) can be asymptotically achieved if the graph $\mathscr{G}$*

has a directed spanning tree, and there exist real matrices $\tilde{P} > 0$, $\tilde{Q} > 0$, $\tilde{R} > 0$, $\tilde{Y} > 0$ and some matrices $\tilde{M}_k$ $(k = 1, 2)$ of appropriate dimensions such that

$$\begin{bmatrix} \tilde{\Gamma}_{11} + h\tilde{\Xi}_{11} & h\tilde{F}^T & h\tilde{M}_1^T \\ * & h(\tilde{\Gamma}_{22} + \mu^2\tilde{Y}) & 0 \\ * & * & -h\tilde{R} \end{bmatrix} < 0 \tag{3.20}$$

$$\begin{bmatrix} \tilde{\Gamma}_{11} & h\tilde{F}^T & h\tilde{M}_2^T \\ * & h\tilde{\Gamma}_{22} & 0 \\ * & * & -h(\tilde{R} + \tilde{Y}) \end{bmatrix} < 0 \tag{3.21}$$

where

$$\begin{aligned} \tilde{\Gamma}_{11} = & \ \mathscr{E}_1^T \tilde{F} + \tilde{F}^T \mathscr{E}_1 - \delta \mathscr{E}_{12}^T (I_{N-1} \otimes \tilde{P}) \mathscr{E}_{12} + \mathscr{E}_1^T \tilde{Q} \mathscr{E}_1 \\ & -\mathscr{E}_3^T \tilde{Q} \mathscr{E}_3 - \mathscr{E}_{23}^T \tilde{M}_1 - \tilde{M}_1^T \mathscr{E}_{23} - \mathscr{E}_{12}^T \tilde{M}_2 - \tilde{M}_2^T \mathscr{E}_{12} \\ & +\gamma \mathscr{E}_{24}^T (E_2^T L^T \Lambda L E_2 \otimes \tilde{\Phi}) \mathscr{E}_{24} - \mathscr{E}_4^T (I_{N-1} \otimes \tilde{\Phi}) \mathscr{E}_4 \\ \tilde{\Xi}_{11} = & \ \delta(\mathscr{E}_{12}^T \tilde{F} + \tilde{F}^T \mathscr{E}_{12}), \ \tilde{\Gamma}_{22} = \mu^2 \tilde{R} - 2\mu I_{N-1} \otimes \tilde{P}, \\ \tilde{F} = & \ (I_{N-1} \otimes A\tilde{P}) \mathscr{E}_1 - (\mathscr{L} \otimes B\tilde{K}) \mathscr{E}_{24} \end{aligned}$$

with γ is given in Theorem 3.2. Moreover, the consensus controller gain is given by

$$K = \tilde{K}\tilde{P}^{-1}.$$

Proof Let $P = I_{N-1} \otimes U$ for the conditions (3.13) and (3.14) in Theorem 3.2, and define $X = \delta P$, $\tilde{P} = U^{-1}$, $\tilde{Q} = P^{-1} Q P^{-1}$, $\tilde{R} = P^{-1} R P^{-1}$, $\tilde{Y} = P^{-1} Y P^{-1}$, $J = diag\{P^{-1}, P^{-1}, P^{-1}, P^{-1}\}$, $\tilde{\Phi} = U^{-1} \Phi U^{-1}$, $\tilde{M}_k = J M_k J (k = 1, 2)$ and $\tilde{K} = K U^{-1}$. Pre- and post-multiplying both the sides of (3.13) by $diag\{J, (R + Y)^{-1}, P^{-1}\}$, and both the sides of (3.14) by $diag\{J, R^{-1}, P^{-1}\}$, respectively, one can obtain (3.20) and (3.21), where the term $P^{-1}(\tilde{R} + \tilde{Y})^{-1} P^{-1}$ is resolved by the inequality

$$-P R^{-1} P \leq \mu^2 R - 2\mu P.$$

The proof is completed. □

Remark 3.8 Note that in [8] and [9] the consensus controller is required to be given **a priori**, otherwise, these methods may be no longer applicable. Therefore, Theorem 3.3 successfully provides an effective method to design an appropriate consensus controller under the event-triggered transmission strategy (3.2).

3.2.3 Co-design Algorithms

Clearly, from Theorem 3.3 one can see that given event-triggered parameters σ_i, the consensus controller K can be derived by solving a set of LMIs in (3.20) and (3.21). Since the event-triggered transmission strategy (3.2) of agent i is highly dependent on its neighbors' sampled-data as well as its own sampled-data, the selection of parameter σ_i may affect not only the transmission number of agent i but also that of its neighbors. Besides, the consensus control performance of the distributed multi-agent system is closely related to the sampled-data transmission number of each agent. Consequently, it is a difficult task to simultaneously obtain the consensus controller K and all the parameters σ_i and Φ of the event-triggered transmission strategy (3.2).

To accomplish this task, we introduce an average transmission rate J on a finite interval $[0, Th)$ as

$$J(\sigma_1, \sigma_2, \ldots, \sigma_N) \triangleq \frac{1}{T\sum_{i=1}^{N} \mathscr{N}_i} \sum_{k=0}^{T-1} \sum_{i=1}^{N} \mathscr{N}_i \rho_k^i \tag{3.22}$$

where T is the sampling number, $\mathscr{N}_i$ denotes the total sampled-data number of the agent i broadcasted to other agents and

$$\rho_k^i = \begin{cases} 1, & \text{if } x_i(kh) \text{ is transmitted} \\ 0, & \text{otherwise} \end{cases} \tag{3.23}$$

represents the sampled-data transmission signal of agent i at each sampling time kh. Notice from (3.22) that, the smaller J can lead to more energy consumption saving. Therefore, the average transmission rate J can be regarded as a performance index of measuring the abilities of the event-triggered transmission strategy. In fact, there is a tradeoff between the resource utilization and the consensus control performance. Thus, to guarantee the consensus performance, we give an expected average transmission rate J^* for the average transmission rate J on the interval $[0, Th)$. Denote the error absolute value between the average transmission rate J and the expected average transmission rate J^* by

$$\varepsilon = |J - J^*|.$$

Hence, the co-design issue turns to simultaneously determining the parameters σ_i and Φ and the controller gain K in order to minimize ε subject to (3.20) and (3.21). For this purpose, we propose Algorithm 1.

Algorithm 1 : Co-design algorithm

STEP 1 Choose a step increment $\theta_i (i = 1, \cdots, N), h, T, \delta, \mu, J^*$ and a sufficiently large initial ε_0. Set the initial parameters $\Lambda_0 = 0$;

STEP 2 Find the maximum σ_i^M of σ_i by solving LMIs in (3.20) and (3.21);

STEP 3 For $\sigma_1 = \theta_1 : \theta_1 : \sigma_1^M$

⋮

For $\sigma_N = \theta_N : \theta_N : \sigma_N^M$

1) Find feasible solutions Φ and K by solving LMIs in (3.20) and (3.21);
2) Calculate J on [0,Th) and ε;
   If $\varepsilon < \varepsilon_0$
   Update $\varepsilon_0 = \varepsilon$, $\Lambda_0 = \Lambda$, $\Phi_0 = \Phi$, $K_0 = K$;
   End

End

⋮

End

STEP 4 Output ε_0, Λ_0, Φ_0 and K_0

3.3 Simulations

In this section, two examples are given to illustrate the effectiveness of the proposed method.

Example 3.3.1 ([4]) Consider the system of four single-integrator agents described by

$$\dot{x}_i(t) = u_i(t), \ i = 1, 2, 3, 4,$$

whose Laplacian matrix L is given by

$$L = \begin{bmatrix} 1 & -1 & 0 & 0 \\ -1 & 3 & -1 & -1 \\ 0 & -1 & 2 & -1 \\ 0 & -1 & -1 & 2 \end{bmatrix}.$$

Choose a set of the parameters

$$\sigma_1 = 0.016, \ \sigma_2 = 0.018, \ \sigma_3 = 0.009, \ \sigma_4 = 0.012.$$

By Theorem 4 in [4], it is known that the inter-event times are lower bounded by 7.4776×10^{-4} s. However, applying Theorem 3.2 in this chapter with $K = 1$, we obtain that the lower bound on inter-event times is $h = 0.0223$ s, which is much larger than 7.4776×10^{-4} s. With this comparison, one conclude that the event-triggered transmission strategy in this chapter can provide a much larger lower bound on inter-event times than the transmission strategy in [4].

Example 3.3.2 Consider the multi-agent system with four linear agents described by

$$\dot{x}_i(t) = \begin{bmatrix} 0 & 1 \\ 0 & -0.4 \end{bmatrix} x_i(t) + \begin{bmatrix} 0.8 \\ 0.5 \end{bmatrix} u_i(t) \ \ (i = 1, 2, 3, 4)$$

The initial states of the multi-agent system are given as

$$x_1(0) = [0; 1], \ x_2(0) = [0.5; 1.5], \ x_3(0) = [-1; 0], \ x_4(0) = [0.5; -0.5].$$

The communication topology of the multi-agent system is given as a directed graph with a spanning tree in Fig. 3.2. Without loss of generality, all the weights are assumed to be 1.

First, we design the consensus controller under the event-triggered transmission strategy. Set the sampling period $h = 0.06s$ and the event-triggered parameters

$$\sigma_1 = 0.010, \ \sigma_2 = 0.007, \ \sigma_3 = 0.016, \ \sigma_4 = 0.013.$$

By solving LMIs in Theorem 3.3 with $\delta = 1$ and $\mu = 1$, we obtain the controller gain as

$$K = \begin{bmatrix} 0.0997 \ 0.3558 \end{bmatrix}.$$

Figure 3.3 illustrates the broadcasting instants and release intervals of each agent on the interval $[0, 30s)$, where the sampled-data transmission ratios of agents 1–4 are, respectively, 9.80%, 7.80%, 6.60% and 9.00%. It is also seen that the average release time of each agent is 0.6122 s, 0.7692 s, 0.9091 s and 0.6667 s, respectively. These results show that the number of the sampled-data transmission is significantly decreased by using the event-triggered transmission strategy. Then, we depict the state trajectories of the system in Fig. 3.4. From this figure, one can clearly see that all the agents' states can indeed reach consensus, which illustrates the effectiveness of the proposed design method.

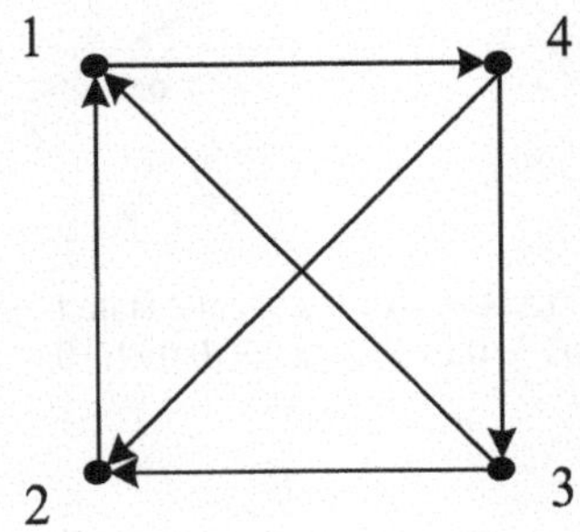

Fig. 3.2 A directed graph $\mathcal{G}$ with a spanning tree

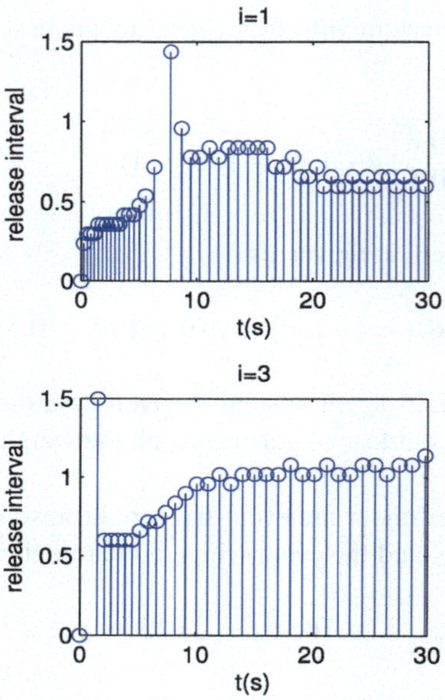

Fig. 3.3 Broadcasting instants and intervals of agent $i(i = 1, 2, 3, 4)$ with $h = 0.06s$, $\sigma_1 = 0.010$, $\sigma_2 = 0.007$, $\sigma_3 = 0.016$, $\sigma_4 = 0.013$ and $K = [0.0997, 0.3558]$

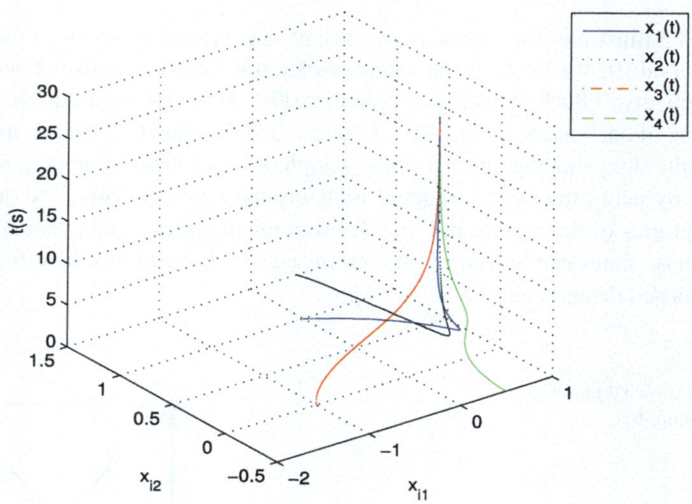

Fig. 3.4 All the agents' state trajectories with $h = 0.06s$, $\sigma_1 = 0.010$, $\sigma_2 = 0.007$, $\sigma_3 = 0.016$, $\sigma_4 = 0.013$ and $K = [0.0997, 0.3558]$

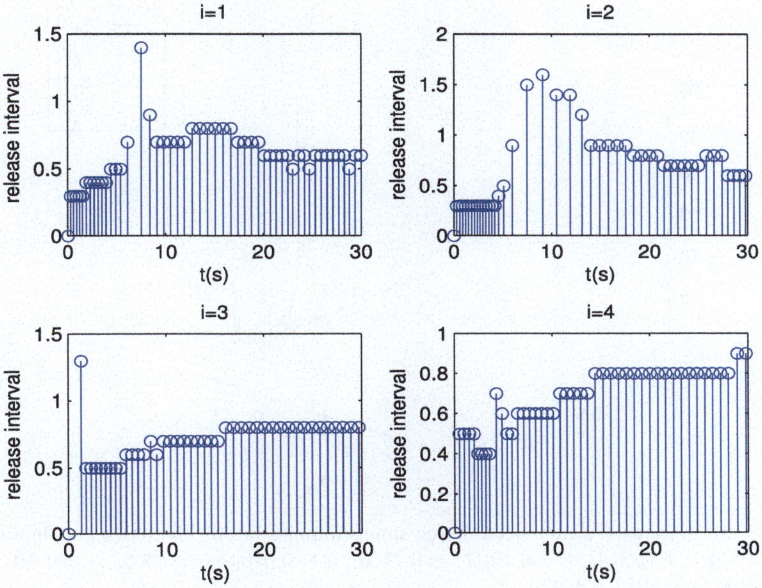

Fig. 3.5 Broadcasting instants and intervals of agent $i(i = 1, 2, 3, 4)$ on simulation interval $[0, 30s)$ based on Algorithm 1: $\theta_i = 0.005$ $(i = 1, 2, 3, 4)$, $h = 0.10$, $J^* = 0.1500$, $\sigma_1 = 0.010$, $\sigma_2 = 0.005$, $\sigma_3 = 0.010$, $\sigma_4 = 0.015$ and $K = [0.1070, 0.3497]$

Second, we co-design the consensus controller gain and the parameters of the event-triggered transmission strategy. Applying Algorithm 1 with $\theta_i = 0.005$ $(i = 1, 2, 3, 4)$, $h = 0.10$, $T = 300$, $\delta = 1$, $\mu = 1$ and $J^* = 0.1500$, we obtain

$$\sigma_1 = 0.010,\ \sigma_2 = 0.005,\ \sigma_3 = 0.010,\ \sigma_4 = 0.015,$$
$$\Phi = \begin{bmatrix} 0.0005 & 0.0017 \\ 0.0017 & 0.0067 \end{bmatrix},\ K = \begin{bmatrix} 0.1070 & 0.3497 \end{bmatrix},\ \varepsilon_{min} = 0.0011$$

and the average transmission rate $J = 0.1511$ which means that only 272 of the total 1800 sampled-data packets are needed to be transmitted during the time period $[0, 30\,\text{s})$. The transmission instants and release intervals of all the agents are illustrated in Fig. 3.5. And Fig. 3.6 plots the all the agents' state trajectories, which implies that the derived results can almost reach the expected average transmission rate J^* while preserving the consensus performance on $[0, 30\,\text{s})$.

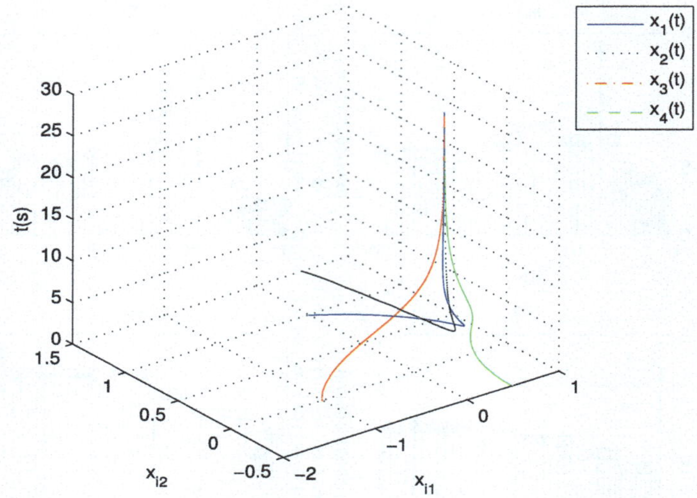

Fig. 3.6 All the agents' state trajectories on simulation interval $[0, 30s)$ based on Algorithm 1: $\theta_i = 0.005$ $(i = 1, 2, 3, 4)$, $h = 0.10$, $J^* = 0.1500$, $\sigma_1 = 0.010$, $\sigma_2 = 0.005$, $\sigma_3 = 0.010$, $\sigma_4 = 0.015$ and $K = [0.1070, 0.3497]$

3.4 Concluding Remarks

A distributed event-triggered transmission strategy for sampled-data consensus of multi-agent systems has been proposed. This strategy only requires the event-triggered condition to be periodically checked and executed at each constant sampling time. Under this strategy, a sampled-data consensus protocol has been developed such that consensus of the distributed multi-agent system can be transformed into the stability of a time-delay system. Then, by employing a Lyapunov-Krasovskii approach, a sufficient condition has been obtained, under which all the states of the distributed multi-agent system asymptotically reach consensus. Correspondingly, based on this stability condition, we have presented a co-design algorithm for determining the parameters of the event-triggered transmission strategy and the consensus controller gain in a unified framework.

3.5 Notes

Compared with [4, 5], one remarkable advantage of the sampled-data-based event-triggered strategy proposed in this chapter is to offer a co-design algorithm of determining the consensus controller gain and the event parameters simultaneously. This feature is very beneficial for practical implementation. Moreover, different from [4, 5], in the sampled-data-based event-triggered setting, it is absolutely guaranteed that the minimum inter-event time is at least one sampling period, which implies that analysis of Zeno behavior is not required any longer. The results in this chapter are based mainly on [10], and the corresponding analysis and design methods can be also referred to [11, 12]. In addition, a variety of other event-triggered schemes for multi-agent systems have been proposed, which can be reviewed in the surveys [13, 14].

References

1. T. Hayakawa, T. Matsuzawa, S. Hara, Formation control of multi-agent systems with sampled information relationship between information exchange structure and control performance, in *Proceedings of IEEE Conference on Decision Control*, pp. 4333–4338 (2006)
2. H. Liu, G. Xie, L. Wang, Necessary and sufficient conditions for solving consensus problems of double-integrator dynamics via sampled control. Int. J. Robust Nonlinear Control **20**(15), 1706–1722 (2010)
3. Y. Cao, W. Ren, Multi-vehicle coordination for double-integrator dynamics under fixed undirected/directed interaction in a sampled-data setting. Int. J. Robust Nonlinear Control **20**(9), 987–1000 (2010)
4. D. Dimarogonas, E. Frazzoli, K. Johansson, Distributed event-triggered control for multi-agent systems. IEEE Trans. Autom. Control **57**(5), 1291–1297 (2012)
5. G. Seyboth, D. Dimarogonas, K. Johansson, Event-based broadcasting for multi-agent average consensus. Automatica **49**, 245–252 (2013)
6. Y.G. Sun, L. Wang, Consensus of multi-agent systems in directed networks with nonuniform time-varying delays. IEEE Trans. Autom. Control **54**(7), 1607–1613 (2009)
7. E. Fridman, A refined input delay approach to sampled-data control. Automatica **46**(2), 421–427 (2010)
8. Z. Li, Z. Duan, G. Chen, L. Huang, Consensus of multiagent systems and synchronization of complex networks: A unified viewpoint. IEEE Trans. Circuits Syst. I: Regul. Pap. **57**(1), 213–224 (2010)
9. K. You, L. Xie, Network topology and communication data rate for consensusability of discrete-time multi-agent systems. IEEE Trans. Autom. Control **56**(10), 2262–2275 (2011)
10. G. Guo, L. Ding, Q.-L. Han, A distributed event-triggered transmission strategy for sampled-data consensus of multi-agent systems. Automatica **50**(5), 1489–1496 (2014)
11. D. Yue, E. Tian, Q.-L. Han, A delay system method for designing event-triggered controllers of networked control systems. IEEE Trans. Autom. Control **58**(2), 475–481 (2013)
12. C. Peng, Q.-L. Han, A novel event-triggered transmission scheme and $\mathscr{L}_2$ control co-design for sampled-data control systems. IEEE Trans. Autom. Control **58**(10), 2620–2626 (2013)
13. L. Ding, Q.-L. Han, X. Ge, X.-M. Zhang, An overview of recent advances in event-triggered consensus of multi-agent systems. IEEE Trans. Cybern. **48**(4), 1110–1123 (2018)
14. X. Ge, Q.-L. Han, L. Ding, Y.-L. Wang, X.-M. Zhang, Dynamic event-triggered distributed coordination control and its applications: a survey of trends and techniques. IEEE Trans. Syst. Man Cybern. Syst. **50**(9), 3112–3124 (2020)

Chapter 4
Distributed Event-Triggered Secondary Control for Islanded Microgrids

This chapter is concerned with active power sharing and frequency regulation in an islanded microgrid under event-triggered communication. A distributed secondary control scheme with a sampled-data-based event-triggered communication mechanism is proposed to achieve active power sharing and frequency regulation in a unified framework, where neighborhood sampled-data exchange occurs only when the predefined triggering condition is violated. Compared with traditional periodic communication mechanisms, the proposed event-triggered communication mechanism shows some prominent ability in reducing the number of communication among neighbors while guaranteeing the desired performance level of microgirds. By employing the Lyapunov-Kravovskii functional method, some sufficient conditions are derived to characterize the effects of control gains, system parameters and sampling period on stability of micgrogrids. Finally, case studies on a modified IEEE 34-bus test system are conducted to evaluate the performance of the proposed distributed control scheme, showcasing its effectiveness, robustness against load changes and plug-and-play ability.

4.1 Problem Formulation

Consider an islanded AC microgrids with N distributed generations (DGs), where each DG unit is composed of a DG voltage source, a voltage source inverter (VSI), an LC filter and an output connector as shown in Fig. 4.1. It is assumed that each DG can communicate with its neighbors via a communication network. A basic hierarchical control framework, which includes primary and secondary control layers, is shown in Fig. 4.1. In what follows, the problem formulation of this chapter is presented based on the control framework.

L. Ding et al., *Distributed Control and Optimization of Networked Microgrids*, Power Systems, https://doi.org/10.1007/978-3-030-95029-3_4

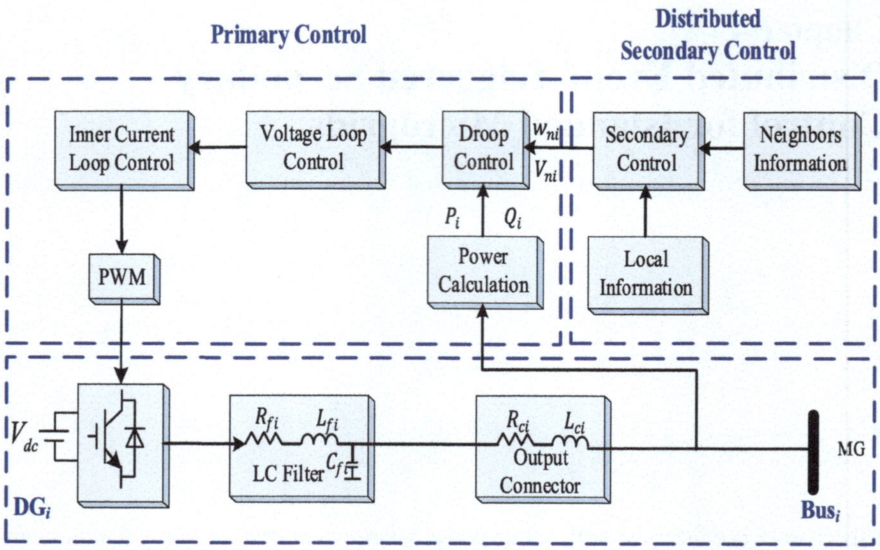

Fig. 4.1 A hierarchical control framework of DG i in an islanded microgrid

Assumption 4.1.1 The communication topology graph $\mathscr{G} = (\mathscr{V}, \mathscr{E})$ is undirected and connected. □

4.1.1 *Primary Droop Control and Secondary Control*

In order to guarantee stability of frequency and voltage while achieving power sharing, a conventional primary droop control (1.1) is first adopted. Notice that the decentralized droop control leads to the deviations of frequency and voltage from their reference values. Thus, a secondary control is introduced to compensate the deviations so that the frequency and voltage regulation can be accurately completed. Taking the derivative of (1.1a) yields

$$\dot{\omega}_i = \dot{\omega}_{ni} - m_i \dot{P}_i = u_i^{\omega} \tag{4.1}$$

when u_i^{ω} is the auxiliary control input of $\dot{\omega}_i$. Then, the nominal set point is determined by

$$\omega_{ni} = \int (u_i^{\omega}(s) + m_i \dot{P}_i(s))ds \tag{4.2}$$

4.1.2 Active Power References

The balance between the total generation active power, load demands and power losses can be described by

$$\sum_{i=1}^{N} P_i = \sum_{i=1}^{N} P_i^L + P_{Loss} \tag{4.3}$$

where P_i^L is the load demand at bus i; P_{Loss} is the total active power loss of the microgrid. According to [1], the power loss is given as $P_{Loss} = \alpha \sum_{i=1}^{N} P_i^L$, where α represents the rate of total load power. Based on the data from energy information administration (EIA) [2], transmission losses of active power can be estimated and calculated, which are approximately 5–7% of total net loads. Thus, the loss rate α is assumed to be a known and small quantity. Consequently, the power balance (4.3) can be rewritten as

$$\sum_{i=1}^{N} P_i = (1+\alpha) \sum_{i=1}^{N} P_i^L \tag{4.4}$$

Since DGs in a microgrid are required to supply the active power to the loads at different buses, it is desirable to realize power sharing in a reasonable and fair way. To deal with this issue, the reference power injection of DG i is set by

$$P_i^{ref} = (1+\alpha)\lambda^d P_i^M \tag{4.5}$$

where $\lambda^d = \frac{\sum_{i=1}^{N} P_i^L}{\sum_{i=1}^{N} P_i^M}$ is the desired utilization level and P_i^M denotes the maximum generation limit of DG i. Suppose that $\sum_{i=1}^{N} P_i^M \geq (1+\alpha) \sum_{i=1}^{N} P_i^L$, which implies that the total power generations are able to meet the total demands under transmission losses without requiring any additional energy storage systems. Under this assumption, it is clear to see that $(1+\alpha)\lambda^d \leq 1$, thus implying that the reference power P_i^{ref} is under the operational capacity limit.

Based on (4.5), we calculate

$$\sum_{i=1}^{N} P_i^{ref} = (1+\alpha)\lambda^d \sum_{i=1}^{N} P_i^M = (1+\alpha) \sum_{i=1}^{N} P_i^L \tag{4.6}$$

which implies that the power balance between supply and demand can be ensured by using the power injection (4.5).

4.1.3 Control Objectives

From the definition of λ^d, one can see that each DG needs to have access to the global information of microgrids including all generation limits P_i^M and load demands P_i^L. In real applications, this requirement may be impractical since parameters of generators and consumers are private and may be unavailable to others. In particular, when the numbers of DGs and loads are large, it becomes costly to obtain all the related information of microgrids. Consequently, it is expected to develop a distributed control algorithm that can converge to the common utilization level λ^d by only using local neighborhood information.

Based on this consideration, a local variable λ_i is introduced for each DG i to estimate the common utilization level λ^d. Then, the reference power P_i^{ref} is modified as

$$P_i^{ref} = (1+\alpha)P_i^M \lambda_i \tag{4.7}$$

Injecting the power (4.7) into (4.2) yields

$$\omega_{ni} = \int (u_i^{\omega}(s) + m_i(1+\alpha)P_i^M u_i^{\lambda}(s))ds$$

where

$$\begin{cases} \dot{\omega}_i = u_i^{\omega} \\ \dot{\lambda}_i = u_i^{\lambda} \end{cases} \tag{4.8}$$

with u_i^{λ} being the auxiliary control input.

Based on the discussion above, the *objective* of our study in this chapter is to design distributed secondary controllers u_i^{ω} and u_i^{λ} in (4.8) such that, under any initials $P_i(0)$ and $\omega_i(0)$,

- The frequency of each DG can be regulated to the reference point, namely,

$$\lim_{t\to\infty} \left|\omega_i(t) - \omega^{ref}\right| = 0. \tag{4.9}$$

  where ω^{ref} is the reference frequency of microgrids;
- The active power sharing can be achieved, equivalently,

$$\lim_{t\to\infty} |\lambda_i(t) - \lambda^d| = 0; \tag{4.10}$$

  while guaranteeing the supply and demand balance (4.4).

4.2 Distributed Event-Triggered Secondary Control

In this section, a sampled-data-based event-triggered communication mechanism is first introduced. Then, with this mechanism, a distributed secondary control scheme for active power sharing and frequency regulation is designed.

4.2.1 Sampled-Data-Based Event-Triggered Communication Schemes

Here, we introduce a sampled-data-based event-triggered communication mechanism to design a distributed secondary control scheme, whose framework is shown in Fig. 4.2. In this framework, an event-triggered communication mechanism is carried out based on periodic sampling with a fixed period $h > 0$. Specifically, an event-triggered module composed of an event detector and a trigger is added to determine when the sampled-data measurement of DGs should be transmitted. The corresponding decision is made by the event-triggered condition (ETC) embedded in the event detector. To make it clear, let the sequence $T_i \triangleq \{t_k^i h \mid k = 0, 1, \ldots, \infty\}$ denote the event-triggered instants with $0 = t_0^i h < t_1^i h < t_2^i h \cdots < t_k^i h < \cdots$. If the predefined

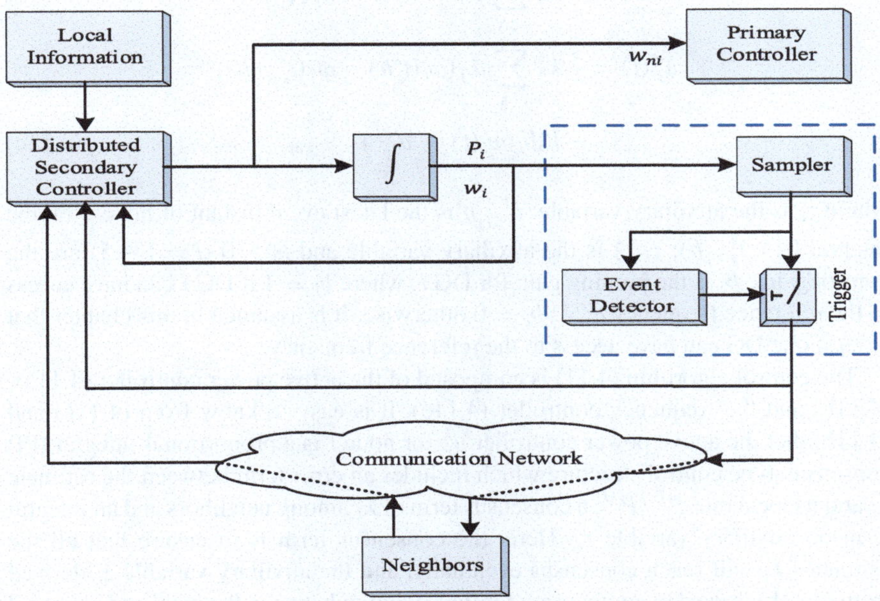

Fig. 4.2 Distributed secondary control structure of DG i with an event-triggered communication mechanism

ETC of DG i is violated at the sampling time instant $t_k^i h$, an event will be immediately triggered, which means that the current sampled-data measurement is permitted to be transmitted to its neighbors through communication networks. In this manner, only the sampled-data measurements at the event-triggered time instants are available to controller updates of neighbors. A zero-order-hold (ZOH) is used to keep the controller update unchanged until the next event-triggered measurement is received.

4.2.2 Distributed Event-Triggered Secondary Controllers

Under event-triggered communication, distributed secondary controllers for active power sharing and frequency regulation are proposed as follows: for $t \in [t_k^i h, t_{k+1}^i h)$

$$
\begin{aligned}
u_i^{\lambda}(t) = & - k_1 (P_i^M \lambda_i(t) - P_i^L) \\
& - k_2 \sum_{j \in N_i} a_{ij} \left(\lambda_i(t_k^i h) - \lambda_j(t_{k(t)}^j h) \right) \\
& + k_3 \sum_{j \in N_i} a_{ij} (z_i(t_k^i h) - z_j(t_{k(t)}^j h) \qquad (4.11a) \\
\dot{z}_i(t) = & - k_4 \sum_{j \in N_i} a_{ij} (\lambda_i(t_k^i h) - \lambda_j(t_{k(t)}^j h)) \qquad (4.11b) \\
u_i^{\omega}(t) = & - k_5 \sum_{j \in N_i} a_{ij} (\omega_i(t_k^i h) - \omega_j(t_{k_j(t)}^j h)) \\
& - k_5 b_i (\omega_i(t) - \omega^{ref}) \qquad (4.11c)
\end{aligned}
$$

where z_i is the auxiliary variable; $t_{k_j(t)}^j h$ is the latest event instant of node j on the interval $[t_k^i h, t_{k+1}^i h)$; $z_i(t)$ is the auxiliary variable and $k_l > 0$ $(l = 1 - 5)$ are the control gains. b_i is the pinning gain for DG i, where $b_i = 1$ if DG i can have access to the reference frequency ω^{ref}; $b_i = 0$ otherwise. It is assumed in this chapter that at least one DG can have access to the reference frequency.

The control algorithm (4.11) is composed of the active power controller (4.11a)-(4.11b) and the frequency controller (4.11c). It is easy to know from (4.11a) and (4.11b) that the active power controller u_i^{λ} for node i is a proportional-integral (PI) consensus-type control structure which includes an error term between the estimate λ_i and its local rate P_i^L / P_i^M, a consensus term of λ_i among neighbors and an integral term on auxiliary variable z_i. Here, the consensus term is to ensure that all the estimates λ_i will reach consensus eventually, and the auxiliary variable z_i derived from (4.11b) is used to compensate for the deviation between the local and expected rates. For the frequency controller (4.11c), a leader-following control structure like

[3] is used, where the reference frequency ω^d can be regarded as a leader. From (4.11), it is observed that only the data $\lambda_j(t^j_{k_j(t)}h)$, $z_j(t^j_{k_j(t)}h)$ and $\omega_j(t^j_{k_j(t)}h)$ of neighbors at the event time $t^j_{k_j(t)}h$ is used rather than $\lambda_j(kh)$ and $z_j(kh)$ and $\omega_j(kh)$ at all sampling instants, thereby leading to less communication between neighbors.

Remark 4.1 Since the operational capacity constraint P^i_{max} on active power of generations is considered in our problem formulation, the transient operation of the secondary control algorithm (4.11a) and (4.11b) may be limited when such maximum generation capacity is reached. Technically, it is certainly possible to deal with this issue by introducing projection operators for DGs [4, 5]. Based on the results in [4, 5], it is easy to know that the projection operation does not affect the convergence of the control algorithm (4.11a) and (4.11b). It is meaningful to further study such an alternative formulation, which will be carefully addressed in our future work.

Remark 4.2 Note that a distributed average consensus algorithm is utilized in [6–10] for power (or load) sharing. Compared with the algorithms in [6–10], the distributed PI consensus-type algorithm in (4.11a) and (4.11b) can bring several remarkable advantages: (1) the algorithms in [6–10] requires information exchange of generations among neighbors such as real-time power measurements, maximum operational power limits or local loads. However, in real energy markets, such individual generation information is always confidential and is not allowed to be accessed by others, which thus restricts implementation of the algorithms in [6, 7]. Differently, the proposed PI consensus-type algorithm only requires to exchange the variables λ_i and z_i rather than any private generation information used by the algorithms of [6–10], thus protecting information privacy of generations; (2) the algorithm in [6] must impose its initials on maximum capacities of generations and local loads. Moreover, since the consensus algorithms in [7–10] converge to the average of initial generation power, the initial power balance between supply and demand is necessary. In contrast, the control algorithm (4.11) is initialization-free and still effective even if there is a mismatch between initial power of total generators and loads.

One principal issue for designing an event-triggered communication mechanism is to select a suitable ETC which seriously affects the number of neighboring communication. In order to address this issue, we define the relative measurement errors between the last triggered and current sampled-data information of DG i as, for $kh \in [t^i_k h, t^i_{k+1}h)$

$$
\begin{aligned}
e^\lambda_i(kh) &\triangleq \lambda_i(t^i_k k) - \lambda_i(kh),\\
e^z_i(kh) &\triangleq z_i(t^i_k k) - z_i(kh),\\
e^\omega_i(kh) &\triangleq \omega_i(t^i_k k) - \omega_i(kh).
\end{aligned}
$$

Then, the triggering functions of DG i are designed as

$$\begin{aligned} f_i^P(kh) &\triangleq g_{1i}|e_i^\lambda(kh)|^2 + g_{2i}|e_i^z(kh)|^2 \\ &\quad - \varepsilon_i \sum_{j\in N_i} a_{ij}\Big(\lambda_i(t_k^i h) - \lambda_j(t_{k_j(t)}^j h)\Big)^2 \\ f_i^\omega(kh) &\triangleq |e_i^\omega(kh)|^2 - \sigma_i\Big[\sum_{j\in N_i} a_{ij}(\omega_i(t_k^i h) - \omega_j(t_{k_j(t)}^j h))^2 \\ &\quad - b_i(\omega_i(kh) - \omega^{ref})^2\Big] \end{aligned}$$

where $g_{1i} > 0$, $g_{2i} > 0$, $\varepsilon_i \geq 0$ and $\sigma_i \geq 0$ are the constants. The event function $f_i^P(kh)$ is used to examine whether the combined variation of $|e_i^\lambda(kh)|$ and $|e_i^z(kh)|$ for the auxiliary variables (λ_i, z_i) exceeds its dynamic threshold $\varepsilon_i \sum_{j\in N_i} a_{ij}(\lambda_i(t_k^i h) - \lambda_j(t_{k_j(t)}^j h))^2$. When $f_i^P(kh) > 0$, it means that the variation is beyond its allowable degree, thereafter the current sampled-data should be transmitted out. Similarly, the event function $f_i^\omega(kh)$ is used to limit the upper bound of frequency variation $|e_i^\omega(kh)|$. As a result, the event instant t_{k+1}^i is determined by the following rule

$$t_{k+1}^i h \triangleq \inf_{kh > t_k^i h} \Big\{ kh \mid f_i^P(kh) > 0 \text{ or } f_i^\omega(kh) > 0 \Big\}. \tag{4.12}$$

It should be pointed out that $e_i^\lambda(kh)$, $e_i^z(kh)$ and $e_i^\omega(kh)$ will be reset as zero after each event time $t_k^i h$.

Remark 4.3 It is worth noting that, sampled data of (λ_i, z_i) and ω_i is encapsulated into a data packet and then transmitted via a commonly shared communication network, hence the communication mechanism (4.12) includes two ETCs of $f_i^P(kh) > 0$ and $f_i^\omega(kh) > 0$, which means that any of excess variations on (λ_i, z_i) and ω_i will trigger the communication mechanism. Technically, such a communication mechanism is easy to install and implement in practical applications. On the other hand, when sampled-data transmission of (λ_i, z_i) and ω_i uses different communication channels or networks, different communication mechanisms may be designed separately for (λ_i, z_i) and ω_i, where the event instants of (λ_i, z_i) are based solely on its ETC of $f_i^P(kh) > 0$ while the event instants of ω_i are based solely on its ETC of $f_i^\omega(kh) > 0$. While such separate mechanisms can improve flexibility and scalability of communication, their costs and complexities in device installation are relatively increased. Therefore, the selection of various communication mechanisms is a specific issue according to practical communication requirements and environments.

Remark 4.4 In comparison to the ETCs in [6, 11–13], the merits of the proposed ETCs in (4.12) lie in: (1) they are checked only at each sampling instant but not continuously, implying that any extra hardware for continuous event monitoring and computing is not needed. Technically, this will be beneficial for practical implementation to reduce costs of system design and maintenance; (2) since the event-triggered communication mechanism (4.12) is on the basis of a periodic sampling scheme, the event instants may happen only at some sampling instants. When the ETCs in (4.12) are satisfied at every sampling time (for example, $\varepsilon_i = 0$ and $\sigma_i = 0$), the event-triggered communication (4.12) will be reduced into a periodic one. In this sense, it is verified that the proposed event-triggered communication mechanism (4.12) always can generate less or at most the same quantities of communication compared with the periodic communication ones; and (3) due to the inherent structure of ETCs in (4.12), its minimum of interevent times is lower bounded by at least one sampling period h, which implies that theoretical analysis on Zeno behavior is no longer required.

4.2.3 Convergence Analysis

To facilitate stability analysis of microgrids with the secondary controller (4.11) under the event-triggered communication mechanism (4.12), we introduce error variables

$$\begin{aligned}
\tilde{\lambda}_i(t) &\triangleq \lambda_i(t) - \lambda^d, \\
\tilde{\omega}_i(t) &\triangleq \omega_i(t) - \omega^{ref}, \\
\tilde{\lambda}(t) &\triangleq \mathrm{col}\{\tilde{\lambda}_1(t), \ldots, \tilde{\lambda}_N(t)\}, \\
z(t) &\triangleq \mathrm{col}\{z_1(t), \ldots, z_N(t)\}, \\
e^{\lambda}(kh) &\triangleq \mathrm{col}\{e_1^{\lambda}(kh), \ldots, e_N^{\lambda}(kh)\}, \\
e^{z}(kh) &\triangleq \mathrm{col}\{e_1^{z}(kh), \ldots, e_N^{z}(kh)\}, \\
\tilde{\omega}(t) &\triangleq \mathrm{col}\{\tilde{\omega}_1(t), \ldots, \tilde{\omega}_N(t)\} \\
e^{\omega}(kh) &\triangleq \mathrm{col}\{e_1^{\omega}(kh), \ldots, e_N^{\omega}(kh)\}.
\end{aligned}$$

Then, for $t \in [kh, kh + h)$, one has

$$\begin{aligned}
\dot{\tilde{\lambda}}(t) =& - k_1(P^M\tilde{\lambda}(t) + P^M\mathbf{1}_N\lambda^d - P^L) - k_2L\tilde{\lambda}(kh) \\
&+ k_3Lz(kh) - k_2Le^{\lambda}(kh) + k_3Le^{z}(kh) \qquad (4.13a) \\
\dot{z}(t) =& - k_4L\tilde{\lambda}(kh) - k_4Le^{\lambda}(kh) \qquad (4.13b) \\
\dot{\tilde{\omega}}(t) =& - B\tilde{\omega}(t) - L\tilde{\omega}(kh) - Le^{\omega}(kh) \qquad (4.13c)
\end{aligned}$$

where $P^M \triangleq \mathrm{diag}\{P_i^M\}$, $P^L \triangleq \mathrm{col}\{P_1^L, \ldots, P_N^L\}$ and $B \triangleq \mathrm{diag}\{b_i\}$.

Define a transformation matrix $S \triangleq \begin{bmatrix} r & R \end{bmatrix}$ [5], where $r \triangleq \frac{1}{\sqrt{N}}\mathbf{1}_N$ and $r^T R = \mathbf{0}_{N-1}^T$ such that $R^T R = I_{N-1}$ and $RR^T = I_N - \frac{1}{N}\mathbf{1}_N\mathbf{1}_N^T$. Then, it is clear that

$$\begin{aligned}
\phi(t) = S^T\tilde{\lambda}(t) &\iff \tilde{\lambda}(t) = S\phi(t), \\
\eta(t) = S^T z(t) - \begin{bmatrix} 0 \\ \Delta \end{bmatrix} &\iff z(t) = S(\eta(t) + \begin{bmatrix} 0 \\ \Delta \end{bmatrix}), \\
\mathscr{E}^{\lambda}(kh) = S^T e^{\lambda}(kh) &\iff e^{\lambda}(kh) = S\mathscr{E}^{\lambda}(kh), \\
\mathscr{E}^{z}(kh) = S^T e^{\lambda}(kh) &\iff e^{\lambda}(kh) = S\mathscr{E}^{z}(kh),
\end{aligned}$$

where $\Delta \triangleq \frac{k_1}{k_3} L_R^{-1} R^T (P^M \mathbf{1}_N \bar{\lambda} - P^L)$ with $L_R \triangleq R^T L R$.

Denote $\phi(t) \triangleq \text{col}\{\phi_1(t), \phi_2(t)\}$ and $\eta(t) \triangleq \text{col}\{\eta_1(t), \eta_2(t)\}$, where $\phi_1(t), \eta_1(t) \in \mathbb{R}$ and $\phi_2(t), \eta_2(t) \in \mathbb{R}^{N-1}$. Under Assumption 4.1.1, it is easy to obtain that

$$L_S \triangleq S^T L S = \begin{bmatrix} 0 & \mathbf{0}_{1\times(N-1)} \\ \mathbf{0}_{(N-1)\times 1} & L_R \end{bmatrix}$$

.

Let $\tau(t) \triangleq t - kh$ for $t \in [kh, kh+h)$, where $0 \leq \tau(t) \leq h$. Then, the system (4.13) can be transformed into

$$\begin{aligned}
\dot{\phi}(t) =& -k_1 P_S^M \phi(t) - k_2 L_S \phi(t-\tau(t)) + k_3 \tilde{L}_R \eta_2(t-\tau(t)) \\
& - k_2 L_S \mathscr{E}^{\lambda}(t-\tau(t)) + k_3 L_S \mathscr{E}^{z}(t-\tau(t)) \quad (4.14a) \\
\dot{\eta}_1(t) =& 0 \quad (4.14b) \\
\dot{\eta}_2(t) =& -k_4 \tilde{L}_R^T \phi(t-\tau(t)) - k_4 \tilde{L}_R^T \mathscr{E}^{\lambda}(t-\tau(t)) \quad (4.14c)
\end{aligned}$$

where $P_S^M \triangleq S^T P^M S$ and $\tilde{L}_R \triangleq [\mathbf{0}_{(N-1)\times 1}, L_R]^T$.

Denoting $y(t) \triangleq \text{col}\{\phi(t), \eta_2(t)\}$ and $\mathscr{E}(t) \triangleq \text{col}\{\mathscr{E}^{\lambda}(t), \mathscr{E}^{z}(t)\}$, the stability of the closed-loop error system (4.14) is equivalent to that of the following delayed systems:

$$\begin{aligned}
\dot{y}(t) =& \mathscr{A}_0 y(t) + \mathscr{A}_1 y(t-\tau(t)) + \mathscr{A}_2 \mathscr{E}(t-\tau(t)) \quad (4.15a) \\
\dot{\tilde{\omega}}(t) =& -B\tilde{\omega}(t) - L\tilde{\omega}(t-\tau(t)) - Le^{\omega}(t-\tau(t)) \quad (4.15b)
\end{aligned}$$

where

$$\begin{aligned}
\mathscr{A}_0 &\triangleq \begin{bmatrix} -k_1 S^T P^M S & \mathbf{0}_{N\times(N-1)} \\ \mathbf{0}_{(N-1)\times N} & \mathbf{0}_{(N-1)\times(N-1)} \end{bmatrix} \\
\mathscr{A}_1 &\triangleq \begin{bmatrix} -k_2 S^T L S & k_3 \tilde{L}_R \\ -k_4 \tilde{L}_R^T & \mathbf{0}_{(N-1)\times(N-1)} \end{bmatrix} \\
\mathscr{A}_2 &\triangleq \begin{bmatrix} -k_2 S^T L S & k_3 S^T L S \\ -k_4 \tilde{L}_R^T & \mathbf{0}_{(N-1)\times N} \end{bmatrix}
\end{aligned}$$

with $\tilde{L}_R \triangleq [\mathbf{0}_{(N-1)\times 1} \quad R^T L R]^T$. For brevity of representation, let $\mathscr{I}_i$ be a block entry matrix with $\mathscr{I}_{ij} = \mathscr{I}_i - \mathscr{I}_j$. For example, $\mathscr{I}_2 = [0, I_{(N-1)n}, 0, 0]$ and $\mathscr{I}_{24} = [0, I_{(N-1)n}, 0, -I_{(N-1)n}]$

To proceed with, an improved reciprocally convex inequality is introduced.

Lemma 4.1 ([14]) *Let $Z_1, Z_2 \in \mathbb{R}^{m\times m}$ be real symmetric positive definite matrices and $\varphi_1, \varphi_2 \in \mathbb{R}^m$. Denote*

$$\mathscr{F}(a) := \frac{1}{a}\varphi_1^T Z_1 \varphi_1 + \frac{1}{1-a}\varphi_2^T Z_2 \varphi_2,$$

where $a \in (0, 1)$. Then, for any $X_1, X_2 \in \mathbb{R}^{m\times m}$, the following inequality holds:

$$\begin{aligned}\mathscr{F}(a) \geq\ & \varphi_1^T [Z_1 + (1-a)(Z_1 - X_1 Z_2^{-1} X_1^T)]\varphi_1 \\ & + \varphi_2^T [Z_2 + a(Z_2 - X_2 Z_1^{-1} X_2^T)]\varphi_2 \\ & + 2\varphi_1^T [a X_1 + (1-a) X_2]\varphi_2\end{aligned}$$

In the sequel, we state and establish the following result.

Theorem 4.1 *Under Assumption 4.1.1, for given scalars $k_l > 0$ ($l = 1, 2, 3, 4, 5$), and any initial conditions $\lambda_i(0)$ and $P_i(0)$ ($i \in \mathscr{V}$), the distributed secondary controller (4.11) with the event-triggered mechanism (4.12) can realize the active power sharing (4.10) and the frequency regulation (4.9) simultaneously, if there exist matrices $Q_1 > 0$, $Q_2 > 0$, $Z > 0$, and real matrices X_1 and X_2 of appropriate dimensions such that*

$$\Phi_p(0) < 0, \quad \Phi_p(1) < 0, \quad p = 1, 2 \tag{4.16}$$

where

$$\Phi_p(a) \triangleq \begin{bmatrix} \Upsilon_p(a, \Theta_p) & F_p^T(a) \\ * & -Z \end{bmatrix}$$

with

$$\begin{aligned}\Upsilon_p(a, \Theta_p) \triangleq\ & J_p^T(a) Q_1 \mathscr{I}_1 + \mathscr{I}_1^T Q_1 J_p(a) + \mathscr{I}_1^T Q_2 \mathscr{I}_1 \\ & - \mathscr{I}_3^T Q_2 \mathscr{I}_3 + [\mathscr{I}_2\ \mathscr{I}_4]\Theta_p[\mathscr{I}_2\ \mathscr{I}_4]^T \\ & - (2-a)\mathscr{I}_{12}^T Z \mathscr{I}_{12} - (1+a)\mathscr{I}_{23}^T Z \mathscr{I}_{23} \\ & - 2\mathscr{I}_{12}^T [a X_1 + (1-a) X_2]\mathscr{I}_{23}, \\ J_1 =\ & \mathscr{A}_0 \mathscr{I}_1 + \mathscr{A}_1 \mathscr{I}_2 + \mathscr{A}_2 \mathscr{I}_4, \\ J_2 =\ & -B\mathscr{I}_1 - L\mathscr{I}_2 - L\mathscr{I}_4, \\ F_p(a) \triangleq\ & hZJ_p + (1-a)X_1^T \mathscr{I}_{12} + a X_2^T \mathscr{I}_{23},\end{aligned}$$

$$\Theta_1 \triangleq \begin{bmatrix} \mathrm{diag}\{S^T LS, \boldsymbol{0}_{N-1}\} & \mathrm{diag}\{S^T LS, \boldsymbol{0}_{(N-1)\times N}\} \\ * & \mathrm{diag}\{S^T LS, \boldsymbol{0}_N\} - G \end{bmatrix},$$

$$\Theta_2 \triangleq \begin{bmatrix} B+L & L \\ L & L - \mathrm{diag}\{1/\sigma_i\} \end{bmatrix},$$

$$G \triangleq \mathrm{diag}\{g_{1i}/\varepsilon_i, g_{2i}/\varepsilon_i\}.$$

Proof Construct the Lyapunov-Krasovskii functional as

$$\begin{aligned} \mathbf{V}(t) =& y^T(t)Q_1y(t) + \int_{t-h}^{t} y^T(s)Q_2y(s)ds \\ &+ h\int_{-h}^{0}\int_{t-\rho}^{t} \dot{y}^T(s)Z\dot{y}(s)dsd\rho \end{aligned} \tag{4.17}$$

Taking the derivative of $\mathbf{V}(t)$ along with (4.14) yields

$$\begin{aligned} \dot{\mathbf{V}}(t) =& 2y^T(t)Q_1\dot{y}(t) + y^T(t)Q_2y(t) + h^2\dot{y}^T(t)Z\dot{y}(t) \\ &- y^T(t-h)Q_2y(t-h) - \mathscr{I}(t) \end{aligned} \tag{4.18}$$

where

$$\mathscr{I}(t) \triangleq h\int_{t-\tau(t)}^{t} \dot{y}^T(s)Z\dot{y}(s)ds + h\int_{t-h}^{t-\tau(t)} \dot{y}^T(s)Z\dot{y}(s)ds.$$

Apply Jensen inequality and Lemma 4.1 to obtain

$$\mathscr{I}(t) \geq \Psi(t)\mathscr{F}_1(a)\Psi(t) \tag{4.19}$$

where $\Psi(t) \triangleq \mathrm{col}\{y(t), y(t-\tau(t)), y(t-h), \mathscr{E}(t-\tau(t))\}$ and

$$\begin{aligned} \mathscr{F}_1(a) \triangleq& \mathscr{I}_{12}^T[Z + (1-a)(Z - X_1Z^{-1}X_1^T)]\mathscr{I}_{12} \\ &+ \mathscr{I}_{23}^T[Z + a(Z - X_2Z^{-1}X_2^T)]\mathscr{I}_{23} \\ &+ 2\mathscr{I}_{12}^T[aX_1 + (1-a)X_2]\mathscr{I}_{23} \end{aligned}$$

with $a = \tau(t)/h$. From the event-triggered condition (4.12), it is easy to obtain that, for $t \in [kh, kh+h)$,

$$g_{1i}|e_i^\lambda(kh)|^2 + g_{2i}|e_i^z(kh)|^2 \leq \varepsilon_i \sum_{j\in N_i} \left(\lambda_i(t_k^i h) - \lambda_j(t_{k_j(t)}^j h)\right)^2$$

which implies that

$$(e^{\lambda}(kh))^T G_1 e^{\lambda}(kh) + (e^{z}(kh))^T G_2 e^{z}(kh)$$
$$\leq \Big(\tilde{\lambda}(kh) + e^{\lambda}(kh)\Big)^T L\Big(\tilde{\lambda}(kh) + e^{\lambda}(kh)\Big) \tag{4.20}$$

where $G_1 = \text{diag}\{g_{1i}/\varepsilon_i\}$ and $G_2 = \text{diag}\{g_{2i}/\varepsilon_i\}$.

Combining (4.18)–(4.20), one has

$$\dot{\mathbf{V}}(t) \leq \Psi^T(t)(\Upsilon_1(a, \Theta_1) + F_1^T(a)ZF_1(a))\Psi(t). \tag{4.21}$$

If $\dot{\mathbf{V}}(t) < 0$, then $y(t)$ is asymptotically stable, which means that the active power sharing can be achieved.

One sufficient condition of $\dot{\mathbf{V}}(t) < 0$ for $\Psi(t) \neq 0$ is that the matrix inequality $\Upsilon_1(a, \Theta_1) + F_1^T(a)ZF_1(a) < 0$ holds. By the Schur complement, $\Upsilon_1(a, \Theta_1) + F_1^T(a)ZF_1(a) < 0$ for $a \in [0, 1]$ if and only if the linear matrix inequalities (LMIs) with $p = 1$ in (4.16) is satisfied. Furthermore, the LMI with $p = 2$ can be obtained by following the same lines of the above proof for the system (4.15b). Therefore, the proof is complete.

Remark 4.5 Theorem 4.1 explicitly characterizes the effects of sampling period, control gains, event parameters, communication topology and system parameters on active power sharing and frequency regulation of microgrids. This is very useful for specific parameter selection in practical implementation. For example, the maximum allowable sampling period h_{max} can be obtained by solving the LMIs in (4.16), thus the selection of sampling period h will be suitable as long as $0 < h \leq h_{max}$.

Remark 4.6 It is well known that the communication topology can directly affect the convergence of distributed algorithms [15]. Note that in this chapter the weighted matrix associated with a communication topology is involved in the LMIs of Theorem 4.1. This means that, for given design parameters, the convergence of the proposed distributed control algorithm can be ensured, provided that the weighted matrix of communication topology is properly chosen to satisfy the LMIs in Theorem 4.1. In this sense, the selection of communication topology plays a critical impact on the convergence of the proposed distributed algorithm.

Remark 4.7 Note that in this chapter, we consider that information of estimates (λ_i, z_i) and frequency ω_i is shared among neighbors via a common communication topology graph $\mathcal{G}$. In practice, however, information transmission of estimates (λ_i, z_i) and frequency ω_i may utilize two different communication topologies (denoted by $\mathcal{G}^P$ and $\mathcal{G}^\omega$, respectively) like [9, 10], in order for security and flexibility of communication. It should be emphasized that, our proposed controller (4.11) can definitely accommodate the case of different topologies $\mathcal{G}^P$ and $\mathcal{G}^w$, provided that both the graphs can satisfy the condition (4.16) in Theorem 4.1. Despite any cases, the proposed distributed control scheme only requires sparse communication networks, leading to low communication burdens, especially in the presence of event-triggered communication mechanisms.

4.2.4 Algorithm Implementation

To implement the distributed secondary control scheme (4.11) with the event-triggered communication mechanism (4.12), Algorithm 2 is given. According to the control framework shown in Fig. 4.1, the proposed distributed secondary control scheme is employed to generate the set point value ω_{ni} for the primary control layer. Then, based on these set-points, the power control loop, the voltage control loop and the current control loop in the primary control layer are executed correspondingly to regulate the output of VSI.

Algorithm 2 Distributed secondary control algorithm considering event-triggered communication

---

Step 1: **Initialize** $\lambda_i(0)$, $z_i(0)$, $\omega_i(0)$ and $t_0^i = 0$; Select a communication topology, a sampling period h, the sampling number T, control gains $k_l(l = 1-5)$ and event parameters g_{1i}, g_{2i} and δ_i;

Step 2: Check the stability condition in Theorem 4.1;

  IF the condition (4.16) is not satisfied
    Return to Step 1 and re-select the parameters;
  ENDIF

Step 3: Update the control algorithm (4.11) based on the event-triggered communication mechanism (4.12);

  WHILE $t = kh(k = 1 : T)$
    Measure the sampled-data $\lambda_i(kh)$, $z_i(kh)$ and $\omega_i(kh)$;
    IF the event condition (4.12) is triggered;
      Transmit the sampled-data states $\lambda_i(kh)$, $z_i(kh)$ and $\omega_i(kh)$ to its neighbors;
      $t_p^i = k$;
    ENDIF
  ENDWHILE

Step 4: Output ω_{ni} and go back to Step 3.

Remark 4.8 It should be mentioned that traditionally, power optimization of microgrids can be realized by addressing an issue of economic dispatch which aims to minimize total generation costs subject to physical constraints of generations. In fact, the power sharing problem considered in this chapter can be regarded as its special case that generation power is proportionally dispatched to loads according to maximum generation capacities. Technically, it is possible to extend our designed control framework to accommodate the generalized economic dispatch for power optimization of microgrids. It is of significance and practice to deeply explore such a problem, which will be further investigated in our future work.

4.3 Case Studies

In order to demonstrate the effectiveness of the proposed distributed secondary control scheme, we consider a modified IEEE 34-bus test system [16], which is shown in Fig. 4.3. In this test system, there are five DGs and three loads connected to the microgrid. It is assumed that each DG can have access to the required information from its neighbors through a communication network. The communication topology of the microgrid is depicted in Fig. 4.4, which is different from its physical topology. According to graph theory, we obtain the adjacency and Laplacian matrices of the communication topology

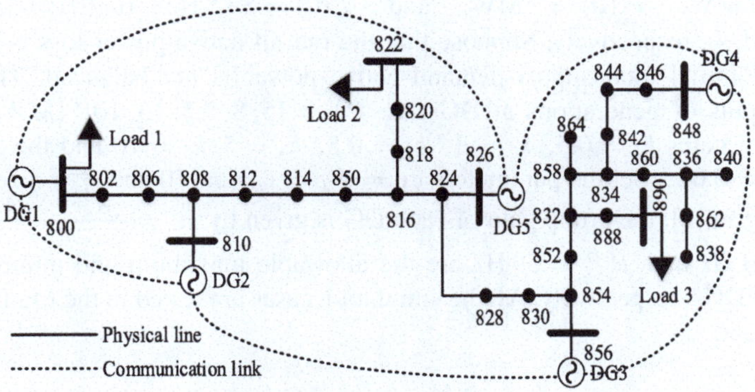

Fig. 4.3 A modified IEEE 34-bus test system

Fig. 4.4 Communication topology

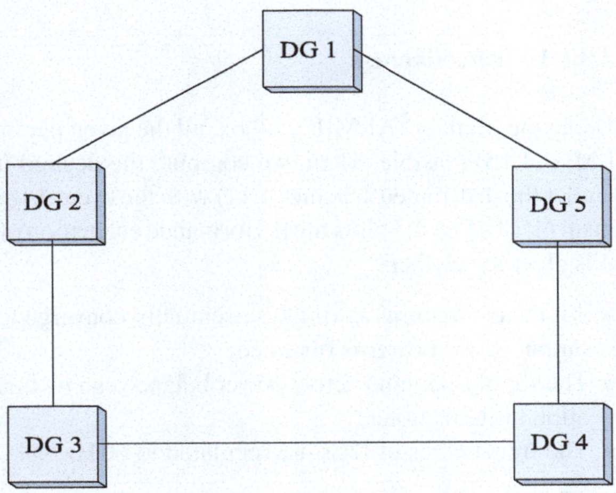

$$A = \begin{bmatrix} 0\,1\,0\,0\,1 \\ 1\,0\,1\,0\,0 \\ 0\,1\,0\,1\,0 \\ 0\,0\,1\,0\,1 \\ 1\,0\,0\,1\,0 \end{bmatrix}, \quad L = \begin{bmatrix} 2 & -1 & 0 & 0 & -1 \\ -1 & 2 & -1 & 0 & 0 \\ 0 & -1 & 2 & -1 & 0 \\ 0 & 0 & -1 & 2 & -1 \\ -1 & 0 & 0 & -1 & 2 \end{bmatrix}.$$

In addition, it is assumed that DG 1 and DG 5 can have access to the reference frequency, i.e. $b_1 = b_5 = 1$.

The parameters of line impedances are referred to [16]. The desired output frequency and voltage of DGs are 60 Hz and 380 V, respectively. The loads are set as $P_1^L = 1.5\,\text{MW}$, $P_2^L = 2.5\,\text{MW}$ and $P_3^L = 2\,\text{MW}$, respectively. The power and frequency initializations of DGs are given by $P_1(0) = P_2(0) = 0.5\,\text{MW}$, $P_3(0) = P_4(0) = 1\,\text{MW}$, $P_5(0) = 2\,\text{MW}$, and $f_i(0) = 59.5\,\text{Hz}$ $(\omega_i(0) = 2\pi f_i(0) \approx 373.85\,\text{rad/s})$, respectively. Suppose that the rate of active power loss is $\alpha = 5\%$. Thus, the initial total supply-demand active power is not balanced. The maximum outputs of generations at DGs are $P^M = [5, 8, 7.5, 15, 10]^T\,(MW)$. Select the control gains $k_1 = 0.5,\ k_2 = 1,\ k_3 = 0.8,\ k_4 = 5,\ k_5 = 10$ and the sampling period $h = 0.02s$. Set the parameters $g_{1i} = g_{2i} = 1/\sigma_i = 10$ and $\varepsilon_i = 1$ for $i \in \mathcal{V}$. According to [1], the droop gain of each DG is given by $m_i = \frac{2\pi(f^{max} - f^{min})}{P_i^M}$, where $f^{max} = 61\,\text{Hz}$ and $f^{min} = 59\,\text{Hz}$ are the allowable maximum and minimum frequency of DGs, respectively. All the simulation cases are tested in the Matlab/PSAT environment.

4.3.1 *Performance Evaluation*

4.3.1.1 Effectiveness

Using the Matlab/YALMIP toolbox, all the given parameters are verified to make the LMIs (4.16) feasible. Then, we compute the desired utilization level $\lambda^d = 0.1319$. Under the distributed scheme (4.11) with the event-triggered communication mechanism (4.12), Fig. 4.5 plots the performance evaluation of the microgrid. From Fig. 4.5, it is clear to see that

- All the estimations λ_i of DGs eventually converge to the desired value λ^d and the output active power is obtained;
- The supply-demand active power balance can be finally realized under any generation initialization;
- The frequencies of DGs are regulated to 60 Hz.

Therefore, the proposed distributed event-triggered control schemes can achieve both active power sharing and frequency regulation by using only the neighborhood data exchange.

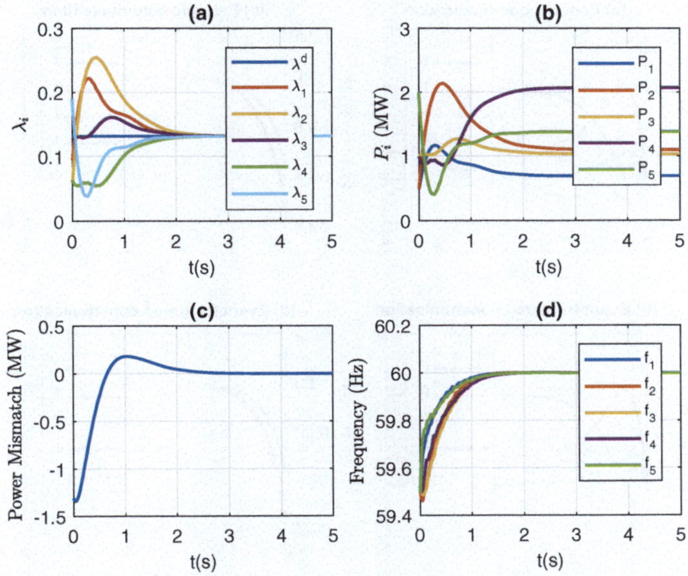

Fig. 4.5 Performance evaluation of distributed controllers (4.11): **a** the trajectories of utilization estimation λ_i; **b** the output active powers; **c** the mismatch of supply and demand powers; **d** the output frequency

4.3.1.2 Comparison to Conventional Periodical Communication Mechanisms

To demonstrate the advantages of the proposed sampled-data even-triggered communication mechanism, a comparison between traditional periodical and event-triggered communication mechanisms is made. The comparison results are shown in Figs. 4.6, 4.7 and 4.8.

It is observed from Fig. 4.6 that the proposed distributed event-triggered control scheme can ensure the nearly identical control performance to traditional periodic control schemes. Under the event-triggered communication mechanisms (4.12), the event instants and data release intervals of DG 5 under periodic and event-triggered communication mechanisms are illustrated in Fig. 4.7, respectively, showing clearly that the information exchange between DG 5 and its neighbors occurs aperiodically. It should be mentioned that, although the power and frequencies of DGs intuitively stay stable after $3s$ in Fig. 4.6, in fact, there still exist slight variations between the actual and reference values. When such slight variations satisfy the ETCs in (4.12), the release intervals may be varying after the time of 3 s.

During the 5 s simulation period, the communication numbers of DGs 1–5 generated by these event-triggered communication mechanisms are 110, 206, 170, 158 and 142, respectively. The number of data transmissions through the communication network is 786, which implies that only 31.31 % (786/2500) sampled-data exchange

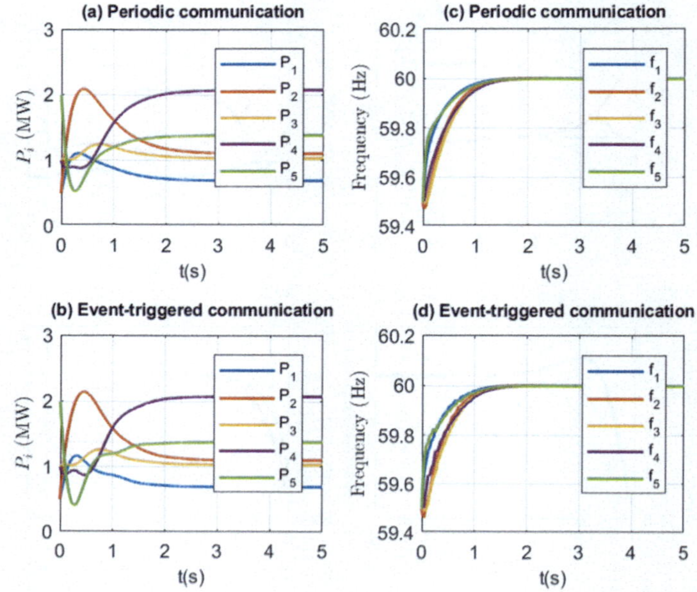

Fig. 4.6 Comparison of control performance under periodic and event-triggered communication

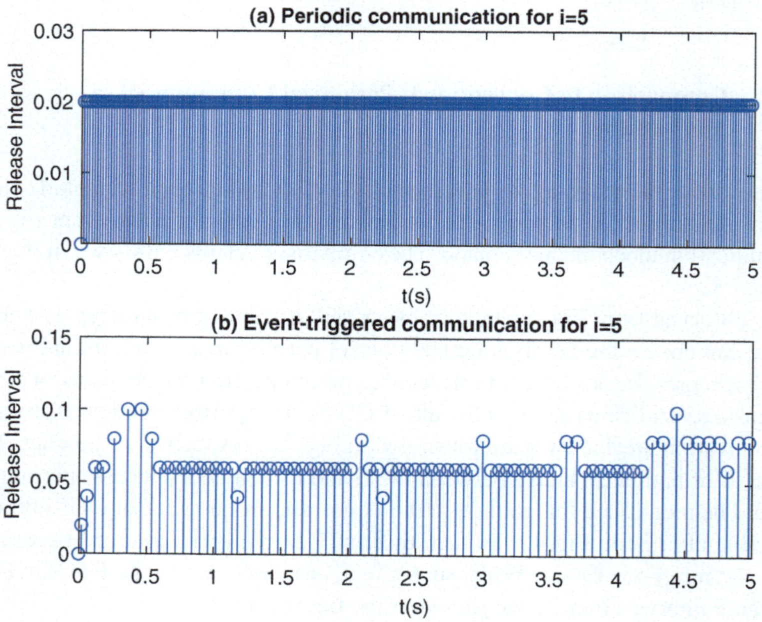

Fig. 4.7 Event instants and intervals of DG 5 under **a** periodic communication; **b** event-triggered communication

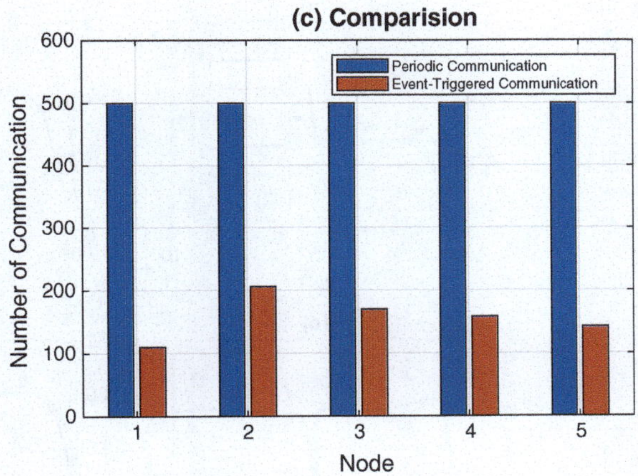

Fig. 4.8 Comparison of total communication number between periodic and event-triggered communications

Table 4.1 Communication rate of DGs with different sampling periods

	DG1	DG2 (%)	DG3 (%)	DG4 (%)	DG5 (%)
h = 0.005	5.19	12.49	10.89	8.59	8.19
h = 0.01	10.18	22.55	20.16	16.37	15.57
h = 0.02	21.91	41.04	33.86	31.47	28.29
h = 0.03	32.21	54.27	47.71	45.33	43.54

is needed. Correspondingly, the comparison of data transmissions under periodic and event-triggered communication mechanisms is depicted in Fig. 4.8. Clearly, event-triggering communication mechanisms (4.12) lead to less consumption of communication than traditional periodic ones.

In order to demonstrate the effects of sampling periods on the proposed event-triggered communication mechanism, the communication rates (transmission number/sampling number) of DGs under event-triggered communication with different sampling periods are listed in Table 4.1. It is observed from Table 4.1 that, when the sampling period h is increased, the communication rates will increase correspondingly. But the communication rates are alway less than or equal to 1 due to the inherent construction characteristic of the proposed event-triggered communication mechanism. In a word, the event-triggered communication mechanism is advantageous over or at least equal to the periodic communication mechanism.

4.3.2 Robustness Against Load Changes

In a microgrid, load demands often suffer from frequent changes according to specific and practical requirements of loads. The robustness of the proposed distributed control scheme against such load changes is evaluated. It is assumed that the demand

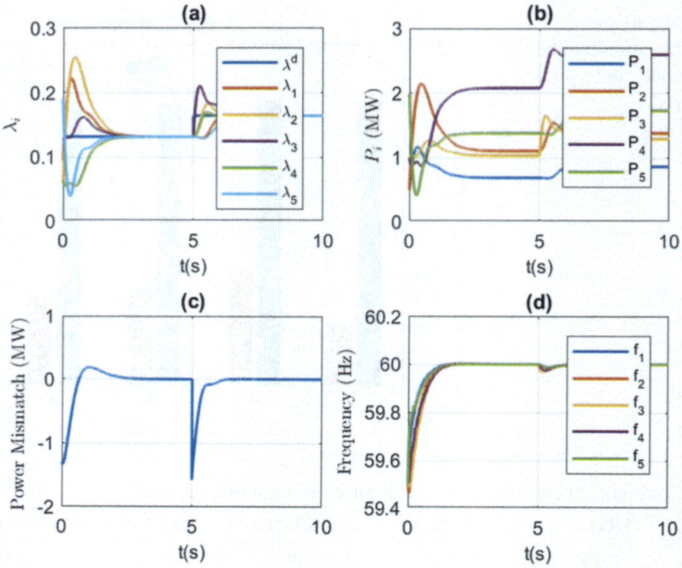

Fig. 4.9 Impacts of load changes: Load 3 is doubled at $t = 5$ s

power of load 3 is doubled at $t = 5s$. Based on the distributed control algorithm (4.11), the impacts of load changes on the microgrid are shown in Fig. 4.9. Recalling (4.5), it is clear that increasing power demand at load 3 yields the higher utilization and the more power generation, which is indeed validated by the simulation results shown in Fig. 4.9a, b. An observation of Fig. 4.9c is that the supply-demand balance is destroyed at $t = 5$ s but restored shortly. From Fig. 4.9d, the frequencies of DGs undergo slight fluctuation caused by load changes at $t = 5$ s, and then they are regulated back to the desired value quickly. Figure 4.10 shows the event instants and release intervals of DG 5. Based on the observation above, it is obvious that the proposed distributed control scheme is of better robustness against load changes by promptly adjusting the power and frequency outputs.

Remark 4.9 It is worth mentioning that load changes in a microgrid result in the reallocation of active power, which can be recognized to as a dynamical economic dispatch problem [17]. In such a dynamical power dispatch, hard constraints on generation capacities and ramping rates of DGs also need to be carefully taken into consideration. According to Remark 1 and Remark 8, it is obvious that the proposed control algorithm can potentially be extended to address such an issue of dynamic economic dispatch under hard constraints. However, this extension of the control algorithm is not presented in the current chapter, which will be further conducted in our future research work.

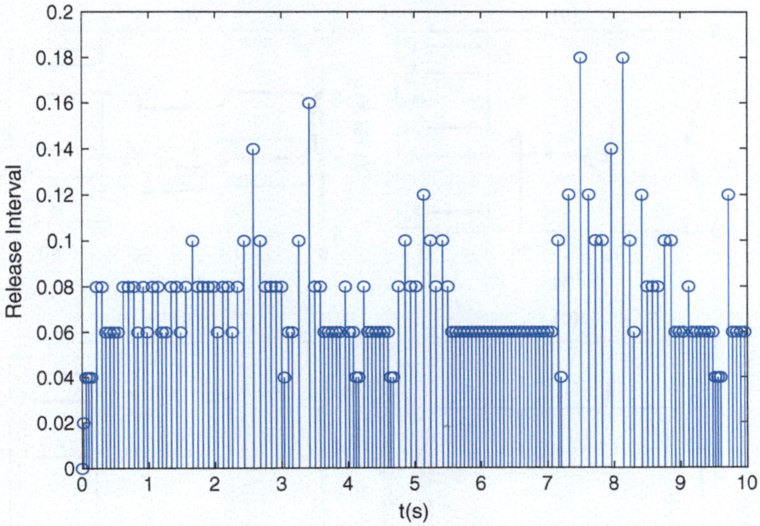

Fig. 4.10 Event instants and intervals of DG 5 under load changes

4.3.3 Plug-and-Play Ability

In a microgrid, the intermittent nature of DGs requires that a good control scheme should provide plug-and-play abilities, namely, it is still effective when some DGs are plugged into or out of the micgrogrid. Moreover, such a plug-and-play operation can guarantee and enhance reliability, flexibility and scalability of microgrids. In what follows, we show that the distributed control schemes (4.11) can realize plug-and-play operation. Suppose that a DG (labeled as DG 6) is added to Bus 838 of the microgrid at $t = 10\,\text{s}$, afterward it is removed again at $t = 20\,\text{s}$. Set the initial power $P_6(0) = 1\,\text{MW}$, the initial frequency $\omega_6(0) = 59.5\,\text{Hz}$ and the maximum power output $P_6^M = 6\,\text{MW}$. During the period $t \in [10\,\text{s}, 20\,\text{s})$, DG 6 can communicate with DG 3. The evolutions of utilization estimation, power generation, and frequency for all the DGs as well as supply-demand power mismatch are shown in Fig. 4.11a. It is observed from Fig. 4.11a, b that the desired utilization level and the reference powers of DGs are decreased for $t \in [10\,\text{s}, 20\,\text{s})$ due to the connection of DG 6 to the microgrid. The event instants and release intervals of DG 5 are shown in Fig. 4.12. It is concluded from Figs. 4.11 and 4.12 that the proposed distributed control scheme under the event-triggered communication mechanism has remarkable plug-and-play abilities.

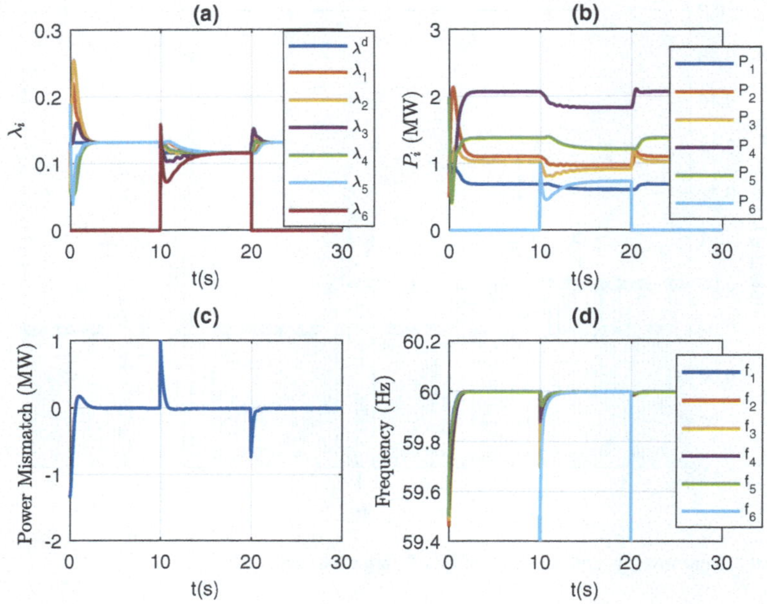

Fig. 4.11 Performance evolution of microgrids under plug-and-play operation

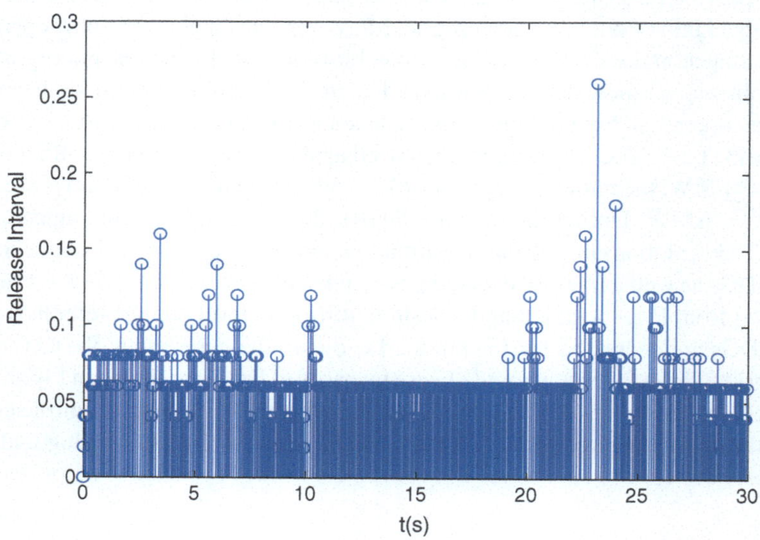

Fig. 4.12 Event instants and intervals of DG 5 under plug-and-play operation

4.4 Concluding Remarks

This chapter has developed a distributed secondary control scheme to realize active power sharing and frequency regulation simultaneously by using only neighborhood information exchange through a communication network. In order to save limited communication resources, an event-triggered communication mechanism has been introduced to determine when the sampled-data of DGs should be transmitted to their neighbors. Some sufficient conditions ensuring accurate active power sharing and frequency regulation have been obtained. By simulation tests of a modified IEEE 34-bus system, it has been verified that the proposed distributed event-triggered control scheme can reduce the communication numbers of DGs significantly while guaranteeing the nearly identical control performance to those of traditional periodic ones.

4.5 Notes

The results in this chapter are based mainly on [18]. It should be mentioned that distributed event-triggered cooperative control frameworks were presented in [6, 11] to realize power sharing, frequency restoration, and voltage control in inverter-based isolated microgrids. However, these event-triggered mechanisms require to monitor and compute event-triggering conditions continuously, which results in much operational burden of algorithm implementation. By contrast, the sampled-data-based event-triggered communication mechanism proposed in this chapter needs to check triggering conditions only at each sampling instant but not continuously, which is easier to implement for practical operation. Note that the proposed control scheme in this chapter requires exact information of loads, which, in practice, may be difficult to accurately measure or estimate. Hence, our future research work will focus on designing a robust event-triggered control scheme regarding load uncertainty. Besides, due to negative impacts of cyber attacks on stability of microgrids, one promising investigation is to explore distributed secondary secure control of islanded microgrids subject to various cyber attacks [19, 20].

References

1. F. Chen, M. Chen, Q. Li, K. Meng, Y. Zheng, J.M. Guerrero, D. Abbott, Cost-based droop schemes for economic dispatch in islanded microgrids. IEEE Trans. Smart Grid **8**(1), 63–74 (2017)
2. L. Wong, A review of transmission losses in planning studies. California Energy Commission, Sacramento, CA, USA, CEC-200-2011-009 (2011)
3. L. Ding, Q.-L. Han, G. Guo, Network-based leader-following consensus for distributed multi-agent systems. Automatica **49**(7), 2281–2286 (2013)

4. L. Ding, Q.-L. Han, L. Wang, E. Sindi, Distributed cooperative optimal control of DC microgrids with communication delays. IEEE Trans. Ind. Inform. **14**(9), 3924–3935 (2018)
5. P. Yi, Y. Hong, F. Liu, Initialization-free distributed algorithms for optimal resource allocation with feasibility constraints and application to economic dispatch of power systems. Automatica **74**, 259–269 (2016)
6. S. Weng, D. Yue, C. Dou, J. Shi, C. Huang, Distributed event-triggered cooperative control for frequency and voltage stability and power sharing in isolated inverter-based microgrid. IEEE Trans. Cybern. **49**(4), 1427–1439 (2018)
7. W. Meng, X. Wang, S. Liu, Distributed load sharing of an inverter-based microgrid with reduced communication. IEEE Trans. Smart Grid **9**(2), 1354–1364 (2018)
8. Y. Xu, H. Sun, W. Gu, Y. Xu, Z. Li, Optimal distributed control for secondary frequency and voltage regulation in an islanded microgrid. IEEE Trans. Ind. Inform. **15**(1), 225–235 (2019)
9. X. Lu, X. Yu, J. Lai, J.M. Guerrero, H. Zhou, Distributed secondary voltage and frequency control for islanded microgrids with uncertain communication links. IEEE Trans. Ind. Inform. **13**(2), 448–460 (2017)
10. X. Lu, X. Yu, J. Lai, J.M. Guerrero, Y. Wang, H. Zhou, A novel distributed secondary coordination control approach for islanded microgrids. IEEE Trans. Smart Grid **9**(4), 2726–2740 (2018)
11. M. Chen, X. Xiao, J.M. Guerrero, Secondary restoration control of islanded microgrids with decentralized event-triggered strategy. IEEE Trans. Ind. Inform. **14**(9), 3870–3880 (2018)
12. Y. Fan, G. Hu, M. Egerstedt, Distributed reactive power sharing control for microgrids with event-triggered communication. IEEE Trans. Control Syst. Technol. **25**(1), 118–128 (2017)
13. L. Ding, L.Y. Wang, G. Yin, W.X. Zheng, Q.-L. Han, Distributed energy management for smart grids with an event-triggered communication scheme. IEEE Trans. Cybern. **27**(5), 1950–1961 (2019)
14. X.-M. Zhang, Q.-L. Han, A. Seuret, F. Gouaisbaut, An improved reciprocally convex inequality and an augmented Lyapunov-Krasovskii functional for stability of linear systems with time-varying delay. Automatica **84**, 221–226 (2017)
15. R. Olfati-Saber, R. Murray, Consensus problems in networks of agents with switching topology and time-delays. IEEE Trans. Autom. Control **49**(9), 1520–1533 (2004)
16. N. Mwakabuta, A. Sekar, Comparative study of the IEEE 34 node test feeder under practical simplifications, in *Proceedings of 39th North American Power Symosium*, Las Cruces, NM. USA (2007), pp. 484–491
17. C. Li, X. Yu, T. Huang, X. He, Distributed optimal consensus over resource allocation network and its application to dynamical economic dispatch. IEEE Trans. Neural Netw. Learn. Syst. **29**(6), 2407–2418 (2018)
18. L. Ding, Q.-L. Han, X.-M. Zhang, Distributed secondary control for active power sharing and frequency regulation in islanded microgrids using an event-triggered communication mechanism. IEEE Trans. Ind. Inform. **15**(7), 3910–3922 (2019)
19. D. Ding, Q.-L. Han, Y. Xiang, X. Ge, X.-M. Zhang, A survey on security control and attack detection for industrial cyber-physical systems. Neurocomputing **275**, 1674–1683 (2018)
20. D. Ding, Z. Wang, Q.-L. Han, G. Wei, Security control for discrete-time stochastic nonlinear systems subject to deception attacks. IEEE Trans. Syst. Man Cybern. Syst. **48**(5), 779–789 (2018)

Chapter 5
Distributed Finite-Time Secondary Control for Islanded Microgrids

This chapter is concerned with distributed secondary frequency and voltage control for islanded microgrids. First, the distributed secondary control problem is formulated by taking both communication delays and switching topologies into account. Second, by using an Artstein model reduction method, a delay-compensated distributed control scheme is proposed to restore frequencies of each distributed generation (DG) to a reference level in finite-time while achieving active power sharing in prescribed finite-time regardless of initial deviations generated from primary control. Third, a distributed finite-time controller is developed to regulate voltages of all DGs to a reference level. Finally, case studies are carried out, demonstrating the effectiveness, the robustness against load changes, and the plug-and-play capability of the proposed controllers.

5.1 Problem Formulation

Consider an islanded AC microgrid with N DGs which can communicate with their neighbors via a communication network. The communication among these DGs is modeled by an undirected graph $\mathscr{G} = (\mathscr{V}, \mathscr{E})$. In a microgrid, each DG unit consists of a DC source, a voltage-controlled voltage source inverter (VCVSI), an LC filter and an RL output connector. Recalling the control structure for DG i in the previous chapter, a block diagram is shown in Fig. 4.1. Similarly, droop control (1.1) is used for the ith DG in the primary control layer. Differentiating (1.1), one obtains

$$\dot{\omega}_i(t) = \dot{\omega}_{ni}(t) - m_i \dot{P}_i(t) = u_{\mathrm{fi}}(t) \tag{5.1}$$

$$\dot{V}_i(t) = \dot{V}_{ni}(t) - n_i \dot{Q}_i(t) = u_{\mathrm{vi}}(t), \tag{5.2}$$

L. Ding et al., *Distributed Control and Optimization of Networked Microgrids*, Power Systems, https://doi.org/10.1007/978-3-030-95029-3_5

where $u_{fi}(t)$ and $u_{vi}(t)$ are frequency and voltage auxiliary controllers, respectively, to be designed. Furthermore, let

$$m_i \dot{P}_i(t) = u_{pi}(t), \tag{5.3}$$

where $u_{pi}(t)$ denotes an active power auxiliary controller to be designed. As a result, $\omega_{ni}(t)$ and $V_{ni}(t)$ are calculated as

$$\omega_{ni}(t) = \int \big(u_{fi}(t) + u_{pi}(t)\big)\mathrm{d}t, \tag{5.4}$$

$$V_{ni}(t) = \int \big(n_i \dot{Q}_i(t) + u_{vi}(t)\big)\mathrm{d}t, \tag{5.5}$$

respectively. Note that a distributed control strategy is advantageous over a centralized one in improving scalability and enhancing reliability of microgrids. Hence, we aim to design distributed secondary controllers $u_{fi}(t)$, $u_{vi}(t)$, and $u_{pi}(t)$ by using neighboring DGs' information through a communication network. When designing a distributed secondary control scheme for islanded microgrids, utilization of communication networks for information exchange among neighboring DGs will introduce the following fundamental issues:

(a) *Communication delays.* Note that communication delays are inherent in a communication system, which are usually caused by limited network bandwidth or congested network traffic. As a result, each DG in a microgrid can only receive the delayed information from its neighbors, and the information transmission during any neighbors may experience different communication delays. Let $h_{ij} > 0$ represent the communication delay when DG i receives its neighbor j's information through a communication network. It is assumed that the communication delay is unknown but upper bounded, namely, $h_{ij} \leq \bar{h}_{ij}$ for $i \in \mathscr{V}$, $j \in \mathscr{N}_i$, where $\bar{h}_{ij}$ is a known positive constant.

(b) *Switching communication topologies.* As stated in Sect. 5.1, due to specific communication requirements or external disturbances, communication topologies of microgrids can switch over time. For example, the original communication topology for DGs is in a pattern as illustrated in the top-left one in Fig. 5.1. Caused by external disturbances, three communication links (between DGs 1 and 3, DGs 2 and 4, DGs 2 and 5) are broken for a while. Then, other backup communication links (between DGs 1 and 2, DGs 3 and 4, DGs 4 and 5) are built, namely, the communication topology is changed to another pattern shown in the top-right one in Fig. 5.1. When other disturbances occur, the topology may switch to another pattern. Therefore, the communication topology can switch among four patterns in Fig. 5.1.

Based on the consideration above, the problem to be addressed is stated as: for the system (5.1)–(5.3) with communication delays and switching communication topologies,

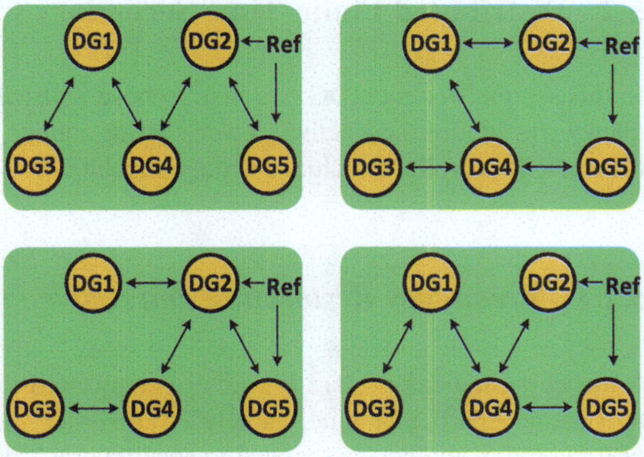

Fig. 5.1 Switching topologies for DGs

(i) design a control set-point $\omega_{ni}(t)$ in (5.4), $i \in \mathscr{V}$, to realize finite-time frequency restoration and prescribed finite-time active power sharing among n DGs, i.e.,

$$\begin{cases} \lim_{t\to T} |\omega_i(t) - \omega_{\text{ref}}| = 0; \\ |\omega_i(t) - \omega_{\text{ref}}| = 0, \forall t \geq T, \forall i \in \mathscr{V}, \end{cases}$$

where ω_{ref} is a reference level of frequency and T is upper bounded by a finite time, and

$$\begin{cases} \lim_{t\to T} |m_i P_i(t) - m_j P_j(t)| = 0; \\ |m_i P_i(t) - m_j P_j(t)| = 0, \forall t \geq T, \forall i, j \in \mathscr{V}, \end{cases}$$

where T is upper bounded by a prescribed finite time that is independent of initial states; and

(ii) design a control set-point $V_{ni}(t)$ in (5.5), $i \in \mathscr{V}$, to achieve finite-time voltage regulation, i.e.,

$$\begin{cases} \lim_{t\to T} |V_i(t) - V_{\text{ref}}| = 0; \\ |V_i(t) - V_{\text{ref}}| = 0, \forall t \geq T, \forall i \in \mathscr{V}, \end{cases}$$

where V_{ref} is the reference level of voltage magnitude and T is upper bounded by a finite time.

Remark 5.1 As mentioned in [1], the line impedance effect makes it difficult to simultaneously fulfill voltage regulation and reactive power sharing except for a perfect and symmetric setup. Therefore, in order to concentrate on the voltage regulation problem, we do not investigate the reactive power sharing problem in this chapter.

5.2 Distributed Finite-Time Secondary Controller Design

In this section, for each DG, we first propose a distributed control scheme for frequency restoration and active power sharing among N DGs. Then, we design a distributed controller to achieve voltage regulation.

5.2.1 Frequency Restoration and Active Power Sharing

A switching indicator $\sigma(t) : \mathbb{R}^+ \rightarrow \Theta$ is employed to determine the graph characteristics at a particular time, namely $\bar{\mathscr{G}}(t) = \bar{\mathscr{G}}_{\sigma(t)}$, where Θ is a set that consists of finite index numbers associated to specific graph patterns. In other words, $a_{ij}(t)$ and $b_i(t)$ are a piecewise constant function containing only finite pieces. In fact, the graph keeps fixed until the next switch occurs when the corresponding graph pattern changes from one to another that is indexed in Θ. The following assumption is made for the switching topologies.

Assumption 5.2.1 For switching graphs $\bar{\mathscr{G}}_{\sigma(t)}$, in each specific graph pattern, the graph corresponding to DGs is connected and the reference levels for frequency and voltage are available to at least one DG. □

Since communication delays are time-varying and heterogeneous, each agent will receive its delayed neighbors' information asynchronously. As a result, this will lead to difficulties in designing finite-time controllers with only past and asynchronously received data information. In order to address this issue, a buffer-based control mechanism is proposed, where each DG can execute control actions after a fixed buffering time so that its controller can use a synchronized data from its own and neighbors. More specifically, in this mechanism, a buffer is introduced for DG i to collect delayed data information from its own and neighbors and store them. As communication delays are unknown but their upper bounds can be known a prior, the buffer length of DG i can be determined by the upper bounds of communication delays. To ensure availability of the past received information for control updates, it is assumed that the buffer for each DG has a length of at least h for each of its neighbors, where $h = \max_{i \in \mathscr{V}} \{\bar{h}_{ij} | j \in \mathscr{N}_i\}$ represents the maximum communication delay in the network. It is clear that the buffer of DG i can store a data set of $N_i \times h$, where N_i is the number of neighbors of DG i. Therefore, the controller of each DG at time t can access the states of its own and its neighbors at as early as the $t - h$ instant from its buffer.

Based on the buffer-based mechanism, the auxiliary controllers for (5.4) are proposed as follows

$$u_{fi}(t) = -\alpha_1 \sum_{j=1}^{N} a_{ij}(t-h)(\xi_i(t-h) - \xi_j(t-h))^{\frac{a}{b}}$$
$$- \alpha_1 b_i(t-h)(\xi_i(t-h) - \omega_{ref})^{\frac{a}{b}}, \tag{5.6}$$
$$u_{pi}(t) = -\alpha_2 \sum_{j=1}^{N} a_{ij}(t-h)(\psi_i(t-h) - \psi_j(t-h))^{\frac{a}{b}}$$
$$- \beta \sum_{j=1}^{N} a_{ij}(t-h)(\psi_i(t-h) - \psi_j(t-h))^{2-\frac{a}{b}}, \tag{5.7}$$

where both a and b are positive odd integers satisfying $a < b$, α_1, α_2 and β are positive control gains, and

$$\xi_i(t-h) = \omega_i(t-h) + \int_{t-h}^{t} u_{fi}(\tau)\mathrm{d}\tau, \tag{5.8}$$
$$\psi_i(t-h) = m_i P_i(t-h) + \int_{t-h}^{t} u_{pi}(\tau)\mathrm{d}\tau. \tag{5.9}$$

It should be pointed out that $(\xi_i(t-h) - \xi_j(t-h))^{\frac{a}{b}}$ in (5.6) is equivalent to[1] $\mathrm{sign}(\xi_i(t-h) - \xi_j(t-h))|(\xi_i(t-h) - \xi_j(t-h))|^{\frac{a}{b}}$ in real domain, which applies to all other exponential calculation in this chapter.

Remark 5.2 To deal with the effects caused from communication delays, two integral compensation terms of $\int_{t-h}^{t} u_{fi}(\tau)\mathrm{d}\tau$ and $\int_{t-h}^{t} u_{pi}(\tau)\mathrm{d}\tau$ are introduced in (5.8)–(5.9), respectively. Note that the delay compensation terms are dependent on the past information of u_{fi} and u_{pi} during the interval $[t-h, t)$. Thus, the buffer of each DG also needs to store its control signals. When the delay compensation terms are removed, the controllers described by (5.6)–(5.9) will be reduced to those commonly used in the literature, thereby the maximum upper bound of delays can be computed. By contrast, as the integral terms can fully compensate the delays, the controllers described by (5.6)–(5.9) are effective even if h is arbitrarily large, which will be theoretically verified in Theorems 5.1 and 5.2.

Let $e_f(t) = [e_{f1}(t), e_{f2}(t), \ldots, e_{fN}(t)]^T$, where $e_{fi}(t) = \xi_i(t) - \omega_{ref}, i \in \mathcal{V}$. From (5.1) and (5.8), the time derivative of $e_{fi}(t)$ is calculated as

$$\dot{e}_{fi}(t) = \dot{\xi}_i(t) = u_{fi}(t) + u_{fi}(t+h) - u_{fi}(t)$$
$$= -\alpha_1 \sum_{j=1}^{N} a_{ij}(t)(e_{fi}(t) - e_{fj}(t))^{\frac{a}{b}} - \alpha_1 b_i(t)(e_{fi}(t))^{\frac{a}{b}}.$$

[1] $\mathrm{sign}(\cdot)$ is the signum function: $\mathrm{sign}(\upsilon)$ equals to 1 if $\upsilon > 0$; equals to -1 if $\upsilon < 0$; and equals to 0 if $\upsilon = 0$.

Construct a Lyapunov function candidate of

$$V_1(t) = \frac{1}{2} e_{\mathrm{f}}^T(t) e_{\mathrm{f}}(t). \tag{5.10}$$

Let $\tilde{\mathscr{G}}(t)$ denote a graph with an adjacency matrix $\tilde{\mathscr{A}}(t) = [a_{ij}^{\frac{2b}{a+b}}(t)]_{N\times N}$ at time t. Let $\tilde{\mathscr{L}}(t)$ denote the Laplacian matrix of graph $\tilde{\mathscr{G}}(t)$. Suppose there exists a diagonal matrix $\tilde{\mathscr{B}}(t) = \mathrm{diag}\{b_1^{\frac{2b}{a+b}}(t), b_2^{\frac{2b}{a+b}}(t), \ldots, b_n^{\frac{2b}{a+b}}(t)\}$, where $b_i(t)$ is the value of b_i at time t. Define a new matrix, $\tilde{\mathscr{M}}(t)$, such that $\tilde{\mathscr{M}}(t) = \tilde{\mathscr{L}}(t) + \tilde{\mathscr{B}}(t)$. Moreover, define $\lambda_{\min}^*(\tilde{\mathscr{M}}) = \min\{\lambda_{\min}(\tilde{\mathscr{M}}(t))\}$, where $\lambda_{\min}(\cdot)$ denotes the smallest eigenvalue of a symmetric matrix.

Let $\hat{\mathscr{G}}(t)$ denote a graph with an adjacency matrix $\hat{\mathscr{A}}(t) = [a_{ij}^{\frac{2b}{3b-a}}(t)]_{N\times N}$ at time t. Let $\hat{\mathscr{L}}(t)$ denote the Laplacian matrix of graph $\hat{\mathscr{G}}(t)$ at time t. Moreover, define $\lambda_2^*(\tilde{\mathscr{L}}) = \min\{\lambda_2(\tilde{\mathscr{L}}(t))\}$ and $\lambda_2^*(\hat{\mathscr{L}}) = \min\{\lambda_2(\hat{\mathscr{L}}(t))\}$, where $\lambda_2(\cdot)$ denotes the second smallest eigenvalue of a symmetric matrix.

We now state and establish the following theorem.

Theorem 5.1 *Under Assumption 5.2.1, for the system described by (5.1) and (5.3), by using controller (5.4) with (5.6)–(5.9), distributed finite-time frequency restoration and prescribed finite-time active power sharing are achieved. Moreover, the upper bounds of the settling time, respectively, are estimated by*

$$T_1 := \frac{2^{\frac{b-a}{b}} b V_1^{\frac{b-a}{2b}}(0)}{\alpha_1(b-a)\lambda_{min}^{*\frac{a+b}{2b}}(\tilde{\mathscr{M}})} + h, \tag{5.11}$$

$$T_2 := \frac{\pi b N^{\frac{b-a}{2b}}}{2(b-a)\lambda_2^{*\frac{a+b}{4b}}(\tilde{\mathscr{L}})\lambda_2^{*\frac{3b-a}{4b}}(\hat{\mathscr{L}})\sqrt{\alpha_2\beta}} + h. \tag{5.12}$$

Proof We first prove that distributed frequency restoration is achieved in finite-time with switching topologies.

From (5.1) to (5.8), the time derivative of $e_{\mathrm{f}i}(t)$ is calculated as

$$\begin{aligned}\dot{e}_{\mathrm{f}i}(t) &= \dot{\xi}_i(t) = u_{\mathrm{f}i}(t) + u_{\mathrm{f}i}(t+h) - u_{\mathrm{f}i}(t)\\ &= -\alpha_1 \sum_{j=1}^{N} a_{ij}(t)(e_{\mathrm{f}i}(t) - e_{\mathrm{f}j}(t))^{\frac{a}{b}} - \alpha_1 b_i(t)(e_{\mathrm{f}i}(t))^{\frac{a}{b}}.\end{aligned}$$

In the sequel, for simplistic presentation, we occasionally drop the dependence on t for the state. At time t, $\sigma(t) \in \Theta$, the time derivative of V_1 is calculated as

$$
\begin{aligned}
\dot{V}_1 =& -\alpha_1 \sum_{i=1}^{N}\sum_{j=1}^{N} a_{ij} e_{\mathrm{f}i}(e_{\mathrm{f}i}-e_{\mathrm{f}j})^{\frac{a}{b}} - \alpha_1 \sum_{i=1}^{N} b_i e_{\mathrm{f}i} e_{\mathrm{f}i}^{\frac{a}{b}} \\
=& -\frac{\alpha_1}{2}\sum_{i=1}^{N}\sum_{j=1}^{N} a_{ij}(e_{\mathrm{f}i}-e_{\mathrm{f}j})^{\frac{a}{b}+1} - \alpha_1 \sum_{i=1}^{N} b_i e_{\mathrm{f}i}^{\frac{a}{b}+1} \\
=& -\frac{\alpha_1}{2}\sum_{i=1}^{N}\sum_{j=1}^{N} \left(a_{ij}^{\frac{2b}{a+b}}(e_{\mathrm{f}i}-e_{\mathrm{f}j})^2\right)^{\frac{a+b}{2b}} \\
& - \alpha_1 \sum_{i=1}^{N} \left(b_i^{\frac{2b}{a+b}} e_{\mathrm{f}i}^2\right)^{\frac{a+b}{2b}} \\
\le& -\frac{\alpha_1}{2}\left(\sum_{i=1}^{N}\sum_{j=1}^{N} a_{ij}^{\frac{2b}{a+b}}(e_{\mathrm{f}i}-e_{\mathrm{f}j})^2 + 2\sum_{i=1}^{N} b_i^{\frac{2b}{a+b}} e_{\mathrm{f}i}^2\right)^{\frac{a+b}{2b}}.
\end{aligned}
$$

Let $\tilde{G}(t) = \sum_{i=1}^{n}\sum_{j=1}^{n} a_{ij}^{\frac{2b}{a+b}}(t)\left(e_{\mathrm{f}i}(t)-e_{\mathrm{f}j}(t)\right)^2 + 2\sum_{i=1}^{n} b_i^{\frac{2b}{a+b}}(t)e_{\mathrm{f}i}^2(t)$. Note that $\tilde{G}(t) = 2e_{\mathrm{f}}^T(t)\tilde{\mathscr{M}}(t)e_{\mathrm{f}}(t)$. Moreover, $\tilde{G}(t) \ge 2\lambda_{\min}(\tilde{\mathscr{M}}(t))e_{\mathrm{f}}^T(t)e_{\mathrm{f}}(t)$. Then, one obtains that $\tilde{G}(t) \ge 4\lambda_{\min}(\tilde{\mathscr{M}}(t))V_1(t)$. Therefore,

$$
\begin{aligned}
\dot{V}_1(t) \le& -\frac{\alpha_1}{2}\left(4\lambda_{\min}(\tilde{\mathscr{M}}(t))V_1(t)\right)^{\frac{a+b}{2b}} \\
=& -2^{\frac{a}{b}}\alpha_1 \lambda_{\min}^{\frac{a+b}{2b}}(\tilde{\mathscr{M}}(t))V_1^{\frac{a+b}{2b}}(t) \\
\le& -2^{\frac{a}{b}}\alpha_1 \lambda_{\min}^{*\frac{a+b}{2b}}(\tilde{\mathscr{M}})V_1^{\frac{a+b}{2b}}(t).
\end{aligned} \tag{5.13}
$$

Applying Theorem 4.2 in [2], one obtains that $V_1(t)$ converges to zero within a finite time of $\check{T}_1 := \frac{2^{\frac{b-a}{b}} b V_1^{\frac{b-a}{2b}}(0)}{\alpha_1(b-a)\lambda_{\min}^{*\frac{a+b}{2b}}(\tilde{\mathscr{M}})}$. So when $t \ge \check{T}_1$, $e_{\mathrm{f}}(t) = 0$. Note that

$$
e_{\mathrm{f}i}(t) = \omega_i(t) - \omega_{\mathrm{ref}} + \int_t^{t+h} u_{\mathrm{f}i}(\tau)\mathrm{d}\tau. \tag{5.14}
$$

Therefore, when $t \ge \check{T}_1$,

$$
\omega_i(t) - \omega_{\mathrm{ref}} = -\int_t^{t+h} u_{\mathrm{f}i}(\tau)\mathrm{d}\tau. \tag{5.15}
$$

Obviously, when $t \ge \check{T}_1 + h$, one obtains that $\omega_i(t) = \omega_{\mathrm{ref}}$. In other words, frequencies of each DG are synchronized to the nominal frequency in a finite time T_1 with switching communication topologies.

Next, we prove that distributed active power sharing is achieved in prescribed finite-time with switching topologies.

Let $e_{\mathrm{p}}(t) = [e_{\mathrm{p}1}(t), e_{\mathrm{p}2}(t), \ldots, e_{\mathrm{p}N}(t)]^T$, where $e_{\mathrm{p}i}(t) = \psi_i(t) - \frac{1}{n}\sum_{i=1}^{N} m_i P_i(t)$, $i \in \mathscr{V}$. From (6) and (16) in the main manuscript, the time derivative of $e_{\mathrm{p}i}(t)$ is calculated as

$$\begin{aligned}
\dot{e}_{\mathrm{p}i}(t) = \dot{\psi}_i(t) &= u_{\mathrm{p}i}(t) + u_{\mathrm{p}i}(t+h) - u_{\mathrm{p}i}(t) \\
&= -\alpha_2 \sum_{j=1}^{N} a_{ij}(t)(\psi_i(t) - \psi_j(t))^{\frac{a}{b}} \\
&\quad - \beta \sum_{j=1}^{N} a_{ij}(t)(\psi_i(t) - \psi_j(t))^{2-\frac{a}{b}} \\
&= -\alpha_2 \sum_{j=1}^{N} a_{ij}(t)(e_{\mathrm{p}i}(t) - e_{\mathrm{p}j}(t))^{\frac{a}{b}} \\
&\quad - \beta \sum_{j=1}^{N} a_{ij}(t)(e_{\mathrm{p}i}(t) - e_{\mathrm{p}j}(t))^{2-\frac{a}{b}}.
\end{aligned} \tag{5.16}$$

Construct a Lyapunov function candidate of

$$V_2 = \frac{1}{2} e_{\mathrm{p}}^T e_{\mathrm{p}}. \tag{5.17}$$

At time t, $\sigma(t) \in \Theta$, the time derivative of V_2 is

$$\begin{aligned}
\dot{V}_2 = &- \alpha_2 \sum_{i=1}^{N}\sum_{j=1}^{N} a_{ij} e_{\mathrm{p}i}(e_{\mathrm{p}i} - e_{\mathrm{p}j})^{\frac{a}{b}} \\
&- \beta \sum_{i=1}^{N}\sum_{j=1}^{N} a_{ij} e_{\mathrm{p}i}(e_{\mathrm{p}i} - e_{\mathrm{p}j})^{2-\frac{a}{b}}
\end{aligned} \tag{5.18}$$

$$\begin{aligned}
= &- \frac{\alpha_2}{2} \sum_{i=1}^{N}\sum_{j=1}^{N} \left(a_{ij}^{\frac{2b}{a+b}} (e_{\mathrm{p}i} - e_{\mathrm{p}j})^2 \right)^{\frac{a+b}{2b}} \\
&- \frac{\beta}{2} \sum_{i=1}^{N}\sum_{j=1}^{N} \left(a_{ij}^{\frac{2b}{3b-a}} (e_{\mathrm{p}i} - e_{\mathrm{p}j})^2 \right)^{\frac{3b-a}{2b}}
\end{aligned} \tag{5.19}$$

$$\begin{aligned}
\le &- \frac{\alpha_2}{2} \left(\sum_{i=1}^{N}\sum_{j=1}^{N} a_{ij}^{\frac{2b}{a+b}} (e_{\mathrm{p}i} - e_{\mathrm{p}j})^2 \right)^{\frac{a+b}{2b}} \\
&- \frac{\beta}{2} N^{\frac{a-b}{b}} \left(\sum_{i=1}^{N}\sum_{j=1}^{N} a_{ij}^{\frac{2b}{3b-a}} (e_{\mathrm{p}i} - e_{\mathrm{p}j})^2 \right)^{\frac{3b-a}{2b}}.
\end{aligned} \tag{5.20}$$

Define

$$\begin{cases} \check{G}(t) = \sum_{i=1}^{N}\sum_{j=1}^{N} a_{ij}^{\frac{2b}{a+b}}(t)\big(e_{\mathrm{p}i}(t) - e_{\mathrm{p}j}(t)\big)^2; \\ \hat{G}(t) = \sum_{i=1}^{N}\sum_{j=1}^{N} a_{ij}^{\frac{2b}{3b-a}}(t)\big(e_{\mathrm{p}i}(t) - e_{\mathrm{p}j}(t)\big)^2. \end{cases}$$

Note that $\check{G}(t) = 2e_{\mathrm{p}}^T(t)\tilde{\mathscr{L}}(t)e_{\mathrm{p}}(t)$ and $\hat{G}(t) = 2e_{\mathrm{p}}^T(t)\hat{\mathscr{L}}(t)e_{\mathrm{p}}(t)$. Moreover, $\check{G}(t) \geq 2\lambda_2(\tilde{\mathscr{L}}(t))e_{\mathrm{q}}^T(t)e_{\mathrm{q}}(t)$, and $\hat{G}(t) \geq 2\lambda_2(\hat{\mathscr{L}}(t))e_{\mathrm{q}}^T(t)e_{\mathrm{q}}(t)$. Then, one obtains that $\check{G}(t) \geq 4\lambda_2(\tilde{\mathscr{L}}(t))V_2(t)$ and $\hat{G}(t) \geq 4\lambda_2(\hat{\mathscr{L}}(t))V_2(t)$. Therefore,

$$\begin{aligned} \dot{V}_2(t) \leq & -\frac{\alpha_2}{2}\left(4\lambda_2(\tilde{\mathscr{L}}(t))V_2(t)\right)^{\frac{a+b}{2b}} \\ & -\frac{\beta}{2}N^{\frac{a-b}{b}}\left(4\lambda_2(\hat{\mathscr{L}}(t))V_2(t)\right)^{\frac{3b-a}{2b}} \\ = & -2^{\frac{a}{b}}\alpha_2\lambda_2^{\frac{a+b}{2b}}(\tilde{\mathscr{L}}(t))V_2^{\frac{a+b}{2b}}(t) \\ & -2^{\frac{2b-a}{b}}\beta N^{\frac{a-b}{b}}\lambda_2^{\frac{3b-a}{2b}}(\hat{\mathscr{L}})(t)V_2^{\frac{3b-a}{2b}}(t) \\ \leq & -2^{\frac{a}{b}}\alpha_2\lambda_2^{*\frac{a+b}{2b}}(\tilde{\mathscr{L}})V_2^{\frac{a+b}{2b}}(t) \\ & -2^{\frac{2b-a}{b}}\beta N^{\frac{a-b}{b}}\lambda_2^{*\frac{3b-a}{2b}}(\hat{\mathscr{L}})V_2^{\frac{3b-a}{2b}}(t). \end{aligned}$$

Applying Lemma 4 in [3], one obtains $V_2(t)$ converges to zero within a prescribed finite time of $\check{T}_2 := \frac{\pi b N^{\frac{b-a}{2b}}}{2(b-a)\lambda_2^{*\frac{a+b}{4b}}(\tilde{\mathscr{L}})\lambda_2^{*\frac{3b-a}{4b}}(\hat{\mathscr{L}})\sqrt{\alpha_2\beta}}$. When $t \geq \check{T}_2$, $e_{\mathrm{p}}(t) = 0$. Note that

$$e_{\mathrm{p}i}(t) = \psi_i(t) - \frac{1}{n}\sum_{i=1}^{n} m_i P_i(t). \tag{5.21}$$

Therefore, when $t \geq \check{T}_2$, $\psi_i(t) = \psi_j(t)$. This in turn implies that $u_{\mathrm{p}i}(t) = 0$ when $t \geq \check{T}_2 + h$. Note that

$$\psi_i(t) = m_i P_i(t) + \int_t^{t+h} u_{\mathrm{p}i}(\tau)\mathrm{d}\tau. \tag{5.22}$$

As a result, when $t \geq \check{T}_2 + h$, one obtains that $m_i P_i(t) = m_j P_j(t)$, which implies that the active power sharing is achieved in a prescribed finite time T_2 with switching communication topologies.

5.2.2 Voltage Regulation

Bearing in mind the buffer-based mechanism, we propose the following auxiliary controller for (5.5)

$$
\begin{aligned}
u_{vi}(t) = & -\alpha_3 \sum_{j=1}^{N} a_{ij}(t-h)(\phi_i(t-h) - \phi_j(t-h))^{\frac{a}{b}} \\
& - \alpha_3 b_i(t-h)(\phi_i(t-h) - V_{\mathrm{ref}})^{\frac{a}{b}},
\end{aligned} \tag{5.23}
$$

where α_3 is a positive control gain, and

$$
\phi_i(t-h) = V_i(t-h) + \int_{t-h}^{t} u_{vi}(\tau)\mathrm{d}\tau. \tag{5.24}
$$

Let $e_{\mathrm{v}}(t) = [e_{\mathrm{v}1}(t), e_{\mathrm{v}2}(t), \ldots, e_{\mathrm{v}N}(t)]^T$, where $e_{\mathrm{v}i}(t) = \phi_i(t) - V_{\mathrm{ref}}$, $i \in \mathscr{V}$. Construct a Lyapunov function candidate of

$$
V_3(t) = \frac{1}{2} e_{\mathrm{v}}^T(t) e_{\mathrm{v}}(t). \tag{5.25}
$$

Then we present the following theorem.

Theorem 5.2 *Under Assumption 5.2.1, for system (5.2), by using controller (5.5) with (5.23)–(5.24), distributed finite-time voltage regulation is achieved with the upper bound of the settling time estimated by*

$$
T_3 := \frac{2^{\frac{b-a}{b}} b V_3^{\frac{b-a}{2b}}(0)}{\alpha_3 (b-a) \lambda_{min}^{*\frac{a+b}{2b}}(\tilde{\mathscr{M}})} + h. \tag{5.26}
$$

Proof The proof follows a similar procedure in Theorem 5.1, so only a schematic proof is provided in what follows. The time derivative of $e_{\mathrm{v}i}(t)$ is calculated as

$$
\begin{aligned}
\dot{e}_{\mathrm{v}i}(t) &= \dot{\phi}_i(t) = u_{\mathrm{v}i}(t+h) \\
&= -\alpha_3 \sum_{j=1}^{N} a_{ij}(t)(e_{\mathrm{v}i}(t) - e_{\mathrm{v}j}(t))^{\frac{a}{b}} - \alpha_3 b_i(t)(e_{\mathrm{v}i}(t))^{\frac{a}{b}}.
\end{aligned}
$$

It can be further calculated that

$$
\begin{aligned}
\dot{V}_3 \leq & -\frac{\alpha_3}{2}\left(\sum_{i=1}^{N}\sum_{j=1}^{N} a_{ij}^{\frac{2b}{a+b}}(e_{\mathrm{v}i} - e_{\mathrm{v}j})^2 + 2\sum_{i=1}^{n} b_i^{\frac{2b}{a+b}} e_{\mathrm{v}i}^2\right)^{\frac{a+b}{2b}} \\
\leq & -2^{\frac{a}{b}} \alpha_3 \lambda_{\min}^{*\frac{a+b}{2b}}(\tilde{\mathscr{M}}) V_3^{\frac{a+b}{2b}}(t).
\end{aligned} \tag{5.27}
$$

Therefore, $V_3(t)$ converges to zero within a finite time of $\check{T}_3 := \frac{2^{\frac{b-a}{b}} b V_3^{\frac{b-a}{2b}}(0)}{\alpha_3(b-a)\lambda_{\min}^{*\frac{a+b}{2b}}(\tilde{\mathscr{M}})}$. So when $t \geq \check{T}_3$, $e_v(t) = 0$, which further indicates that when $t \geq T_3$, $V_i(t) = V_{\text{ref}}$. This completes the proof.

Remark 5.3 Note that Theorems 5.1 and 5.2 can explicitly characterize the effects of communication delays on the convergence speeds of the proposed control schemes which are associated with T_1, T_2 and T_3, respectively. It is clear that all the upper bounds of the settling time T_1, T_2 and T_3 closely rely on h, which implies that T_1, T_2 and T_3 will become larger when h increases.

5.3 Case Studies

In this section, case studies are conducted on a modified IEEE 34-bus test system [4], which is shown in Fig. 5.2. The test system consists of five DGs (i.e., $N = 5$) and three loads. The communication graph is arbitrarily switching among four patterns shown in Fig. 5.1. The information of reference levels for frequency and voltage is available to DG 2 and DG 5. The case studies are performed in the Matlab/PSAT environment.

The reference levels for frequency and voltage 50 Hz and 380 V, respectively. The droop coefficients are set as $m_1 = m_5 = 4.2 \times 10^{-5}$, $m_2 = m_3 = m_4 = 8.4 \times 10^{-5}$. Other used parameters of line impedances are refereed to [4]. Note that we focus on the period after the primary control, and treat $t = 0$s as the starting point of the secondary control. At $t = 0$s, the test system is initialized with $\omega_1(0) = \omega_2(0) = \omega_3(0) = \omega_4(0) = \omega_5(0) = 49.5$ Hz; $P_1(0) = P_5(0) = 50$KW, $P_2(0) = 20$ KW, $P_3(0) = 30$ KW, $P_4(0) = 10$ KW; $V_1(0) = 376.9$ V, $V_2(0) = 377.4$ V, $V_3(0) = 373.8$ V, $V_4(0) = 374.3$ V, $V_5(0) = 379.4$V. The communication delays are set as: $h = 0.36$s. The controller parameters are selected as: $\alpha_1 = 2.5$, $\alpha_2 = 0.4$, $\alpha_3 = 0.9$, $\beta = 3.5$.

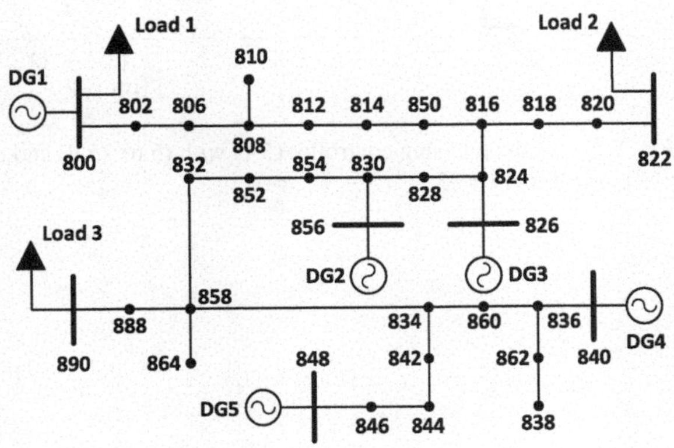

Fig. 5.2 The modified IEEE 34 bus test feeder

5.3.1 Performance Evaluation

Applying controller (5.4) with (5.6)–(5.9) and controller (5.5) with (5.23)–(5.24), for $N = 5$, when $a = 1$ and $b = 5$, results are shown in Fig. 5.3. When setting $a = 13$ and $b = 15$, results are shown in Fig. 5.4. Although the consensus is achieved in a shorter time when $a = 1$ and $b = 5$, there exists more significant oscillation (as shown in the first plot of Fig. 5.3), which is an undesired behavior in practical implementation. Therefore, in the remaining part of Sect. 5.3, we set $a = 13$ and $b = 15$.

In Fig. 5.4, frequency restoration, active power sharing, and voltage regulation are achieved at $t = 5.9$ s, $t = 7.2$ s, and $t = 5.7$ s, respectively. To make a comparison, we show the results by using controllers without delay compensation in Fig. 5.5, where significant fluctuations or even divergences occur. These kind of fluctuations and divergences can result in undesired performance or even lead to the failure of DGs. The comparisons between Figs. 5.4 and 5.5 show that the proposed controller is effective for a micrigrid that suffers from communication delays.

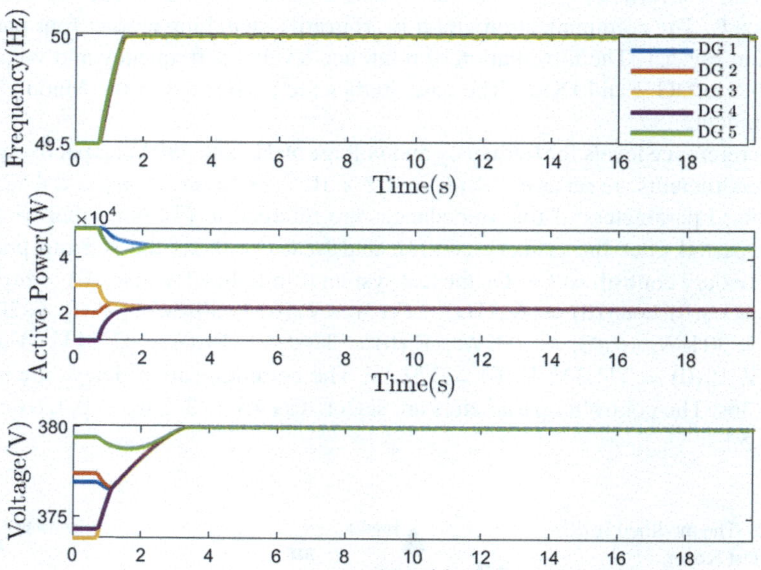

Fig. 5.3 Results by using controller (5.4) with (5.6)–(5.9) and controller (5.5) with (5.23)–(5.24) when $a = 1$ and $b = 5$

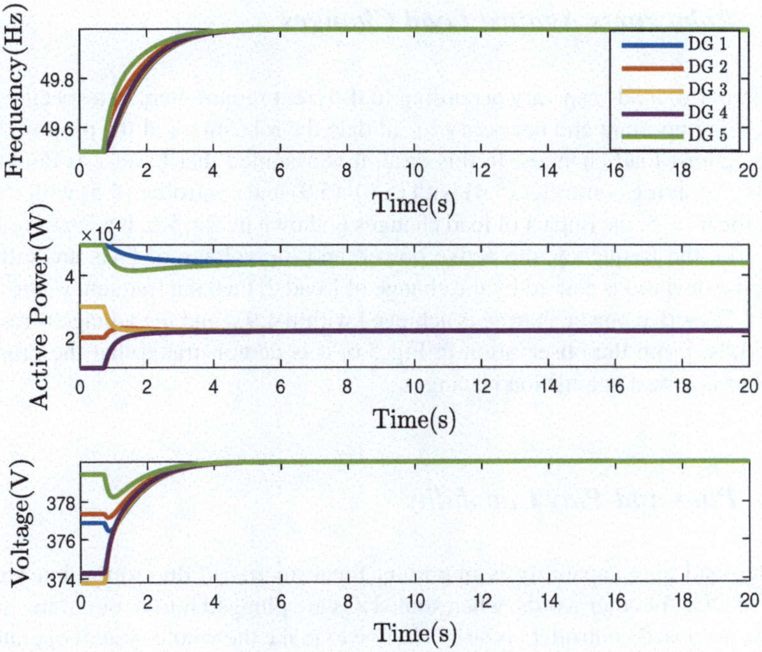

Fig. 5.4 Results by using controller (5.4) with (5.6)–(5.9) and controller (5.5) with (5.23)–(5.24) when $a = 13$ and $b = 15$

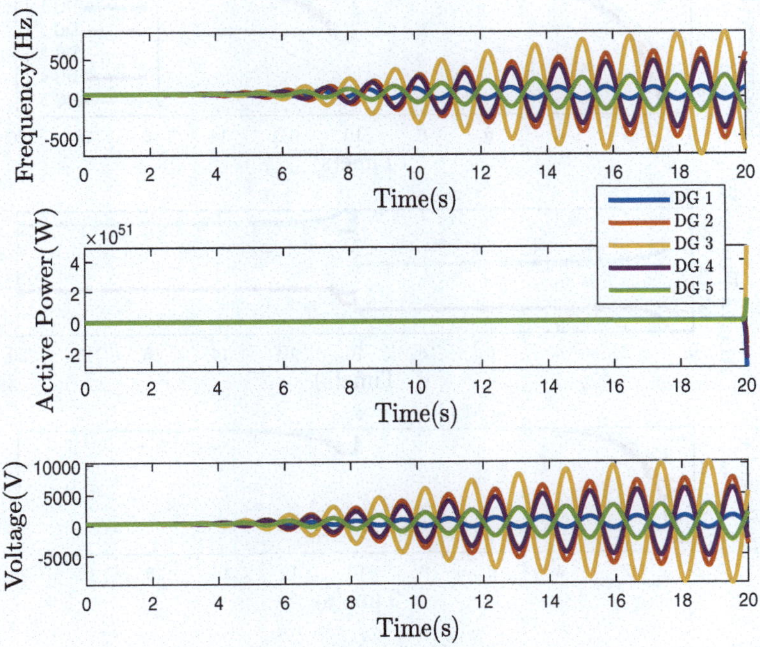

Fig. 5.5 Results without delay compensation

5.3.2 Robustness Against Load Changes

In a microgrid, loads can vary according to different requirement at a specific time. Hence, it is important and necessary to validate the robustness of the proposed controllers against load changes. In this case, it is assumed that Load 2 is doubled at $t = 10$ s. Applying controller (5.4) with (5.6)–(5.9) and controller (5.5) with (5.23)–(5.24), for $N = 5$, the impact of load changes is shown in Fig. 5.6. It is observed that, at $t = 10$ s, the frequency, the active power, and the voltage of DGs are suffering from some deviations caused by the change of Load 2, then the frequency is restored within 4.2s, active power sharing is achieved within 4.9 s, and the voltage is restored within 3.8s. From the observation in Fig. 5.6, it is demonstrated that the proposed controller is robust against load changes.

5.3.3 Plug-and-Play Capability

The plug-and-play capability is important for a microgrid due to the intermittent nature of DGs. In other words, when some DGs are plugged into or out of the microgrid, the proposed controllers is still effective to make the whole system operate in a

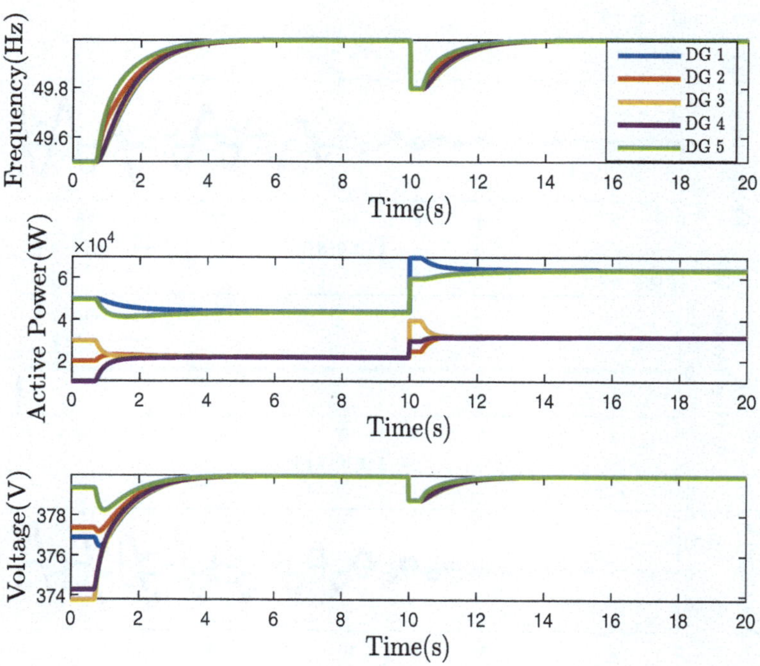

Fig. 5.6 Robustness to load changes (doubled Load 2)

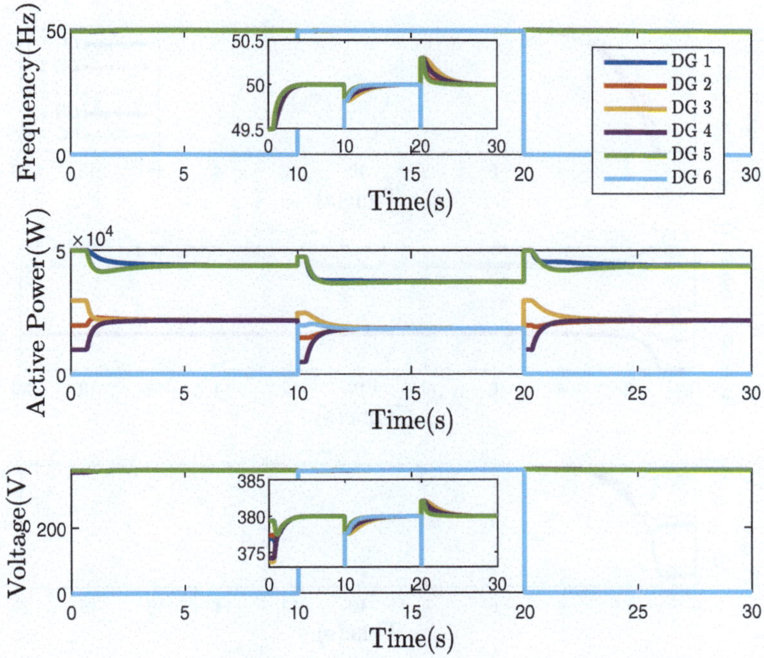

Fig. 5.7 Plug-and-Play capability (DG 6 is added at $t = 10\,\text{s}$ and then is removed at $t = 20\,\text{s}$)

normal mode. To address this concern, it is assumed that a "DG 6", with communication delay $h_6 = 0.36\,\text{s}$, is added to Bus 852 of the microgrid at $t = 10\,\text{s}$, and then is removed at $t = 20\,\text{s}$. Besides, it is considered that DG 6 is a neighbor of DG 5 for $10\,\text{s} \leq t \leq 20\,\text{s}$ on top of the existing topology in Fig. 5.1. Applying controller (5.4) with (5.6)–(5.9) and controller (5.5) with (5.23)–(5.24), the impact of DG 6 is shown in Fig. 5.7. It is observed that, at $t = 10\,\text{s}$, the frequency, the active power, and the voltage of DGs are suffering from some deviations caused by the addition of DG 6, then the frequency is restored within 7.1 s, active power sharing is achieved within 6.1 s, and the voltage is restored within 6.6 s. Once DG 6 is removed from the microgrid, the frequency, the active power, and the voltage again suffer from some deviations within a range, then return to the reference level (for frequency and voltage) and achieve active power sharing.

5.3.4 *Two Different Sets of Communication Delays*

In this section, we consider two different sets of communication delays. Particularly, choose delay set (i): $h = 0.55\,\text{s}$; and delay set (ii): $h = 1.55\,\text{s}$. Corresponding results are presented in Figs. 5.8, 5.9, 5.10 and 5.11. It can be observed that using controllers

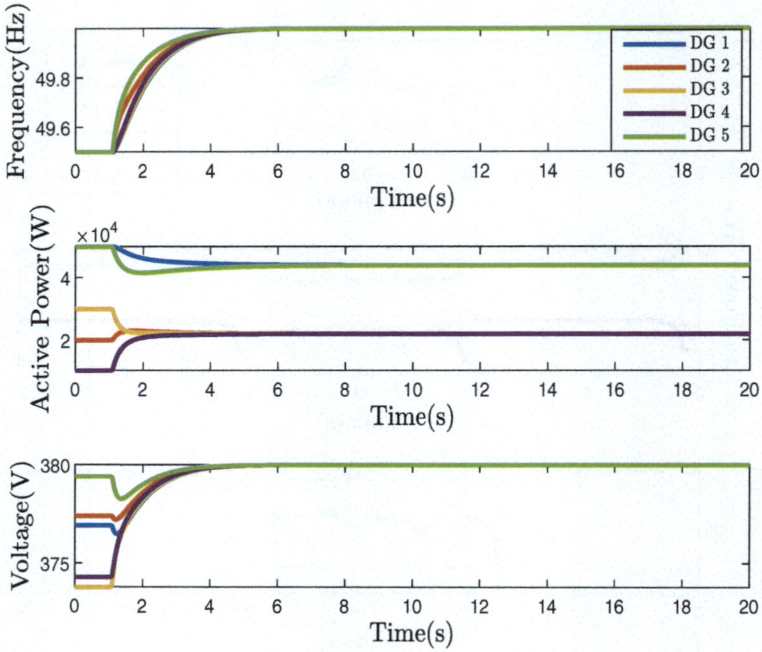

Fig. 5.8 Results with delay compensation under delay set (i)

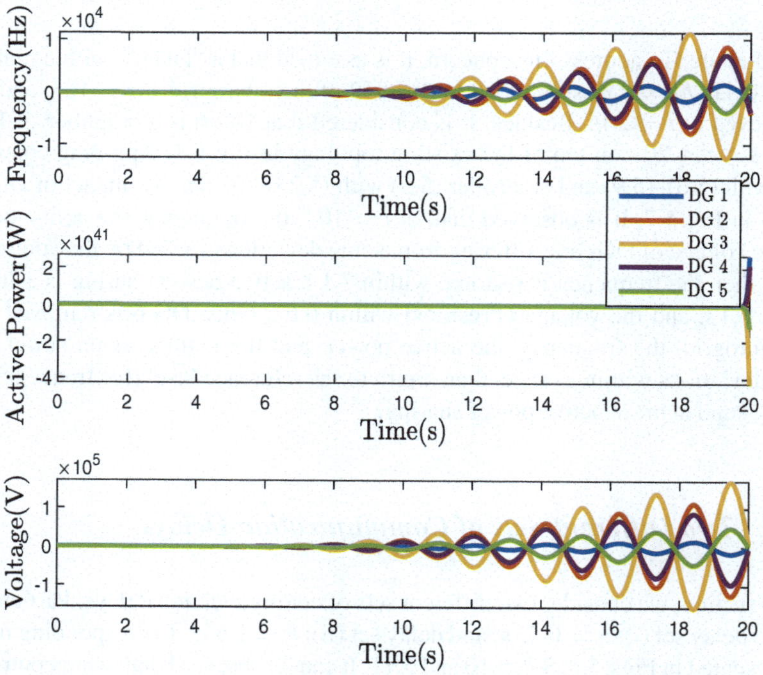

Fig. 5.9 Results without delay compensation under delay set (i)

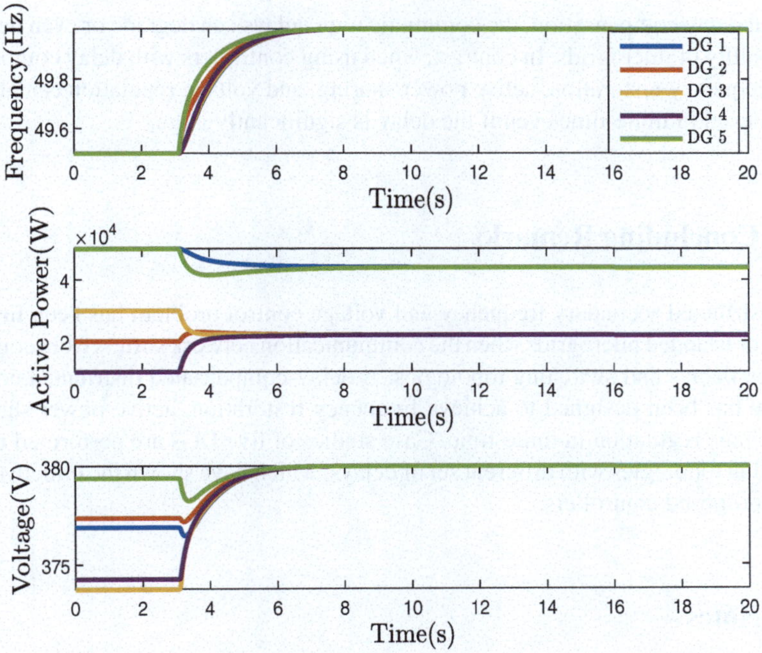

Fig. 5.10 Results with delay compensation under delay set (ii)

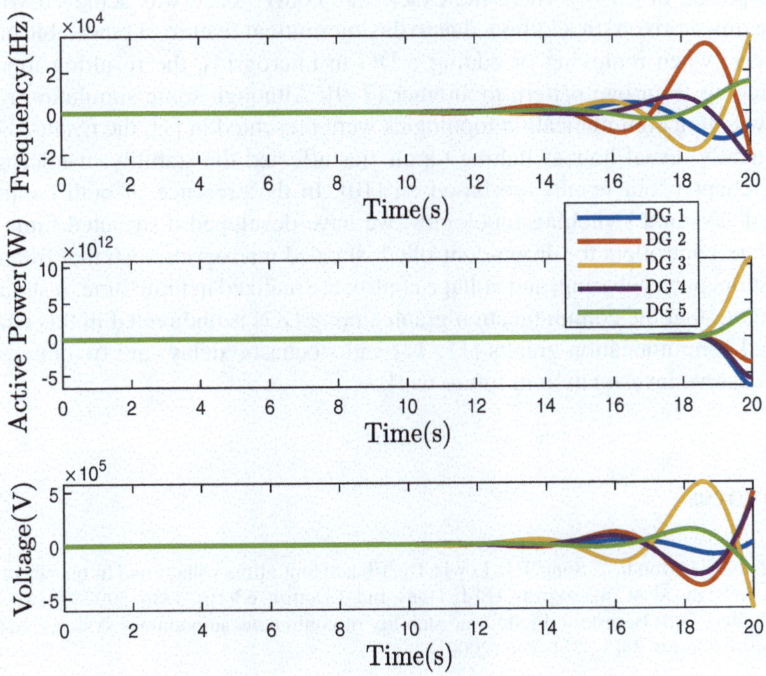

Fig. 5.11 Results without delay compensation under delay set (ii)

without delay compensation, the communication delays can degrade or even destroy the stability of microgrids. In contrast, when using controllers with delay compensation, frequency restoration, active power sharing, and voltage regulation can always be achieved in finite-time even if the delay is significantly large.

5.4 Concluding Remarks

The distributed secondary frequency and voltage control problem has been investigated for islanded microgrids when the communication network suffers from communication delays and switching topologies. A delay-compensated distributed control scheme has been designed to achieve frequency restoration, active power sharing, and voltage regulation in finite-time. Case studies of five DGs are performed under switching topologies with different set of delays, which have shown the effectiveness of the proposed controllers.

5.5 Notes

Some related works regarding the effects of communication delays on microgrid were reported in [5, 6], where the exact state convergence was achieved with an infinite time horizon. In addition, due to the intermittent feature of renewable energy resources, when removing or adding a DG in microgrids, the resulting topology is switching from one pattern to another [7–9]. Although some simulation results with switching communication topologies were presented in [1], the results did not theoretically reveal how switching topologies affected the stability of microgrids. In this chapter, our results are based on [10]. In the presence of both communication delays and switching topologies, we have developed distributed finite-time secondary controllers for droop-controlled islanded microgrids such that frequency restoration, power sharing, and voltage control are realized in finite time. It should be pointed out that the communication graph among DGs is undirected in this chapter. Directed communication graphs [11, 12] and stochastic delays are two interesting topics deserve investigation in future work.

References

1. S. Zuo, A. Davoudi, Y. Song, F.L. Lewis, Distributed finite-time voltage and frequency restoration in islanded AC microgrids. IEEE Trans. Ind. Electron. **63**(10), 5988–5997 (2016)
2. S.P. Bhat, D.S. Bernstein, Finite-time stability of continuous autonomous systems. SIAM J. Control. Optim. **38**(3), 751–766 (2000)

3. S. Parsegov, A. Polyakov, P. Shcherbakov, Fixed-time consensus algorithm for multi-agent systems with integrator dynamics, in *Proceedings of 4th IFAC Workshop Distrbute Estimation Control Network System* (2013), pp. 110–115
4. A. Bidram, A. Davoudi, F.L. Lewis, A multiobjective distributed control framework for islanded AC microgrids. IEEE Trans. Ind. Inform. **10**(3), 785–1798 (2014)
5. C. Dou, D. Yue, J.M. Guerrero, X. Xie, S. Hu, Multiagent system-based distributed coordinated control for radial DC microgrid considering transmission time delays. IEEE Trans. Smart Grid **8**(5), 2370–2381 (2017)
6. L. Ding, Q.-L. Han, L. Wang, E. Sindi, Distributed cooperative optimal control of DC microgrids with communication delays. IEEE Trans. Ind. Inform. **14**(9), 3924–3935 (2018)
7. J. Qin, C. Yu, Coordination of multiagents interacting under independent position and velocity topologies. IEEE Trans. Neural Netw. Learn. Syst. **24**(10), 1588–1597 (2013)
8. Y. Liu, G.-H. Yang, Prescribed performance-based consensus of nonlinear multiagent systems with unknown control directions and switching networks. IEEE Trans. Syst., Man, Cybern., Syst. **50**(2), 609–616 (2020)
9. X. Wang, H. Zhang, C. Li, Distributed finite-time cooperative control of droop-controlled microgrids under switching topology. IET Renew. Power Gener. **11**(5), 707–714 (2017)
10. B. Ning, Q.-L. Han, L. Ding, Distributed edge-based finite-time secondary frequency and voltage control for islanded microgrids with communication delays and switching topologies. IEEE Trans. Cybern. **51**(8), 3988–3999 (2021)
11. H. Li, C. Huang, G. Chen, X. Liao, T. Huang, Distributed consensus optimization in multiagent networks with time-varying directed topologies and quantized communication. IEEE Trans. Cybern. **47**(8), 2044–2057 (2017)
12. H. Li, G. Chen, T. Huang, Z. Dong, High-performance consensus control in networked systems with limited bandwidth communication and time-varying directed topologies. IEEE Trans. Neural Netw. Learn Syst. **28**(5), 1043–1054 (2017)

Chapter 6
Distributed Resilient Finite-Time Secondary Control for Heterogeneous BESSs

This chapter addresses the problem of distributed resilient finite-time control of multiple heterogeneous battery energy storage systems (BESSs) in a microgrid subject to denial of service (DoS) attacks. Notice that, DoS attacks may block information transmission among BESSs by preventing the BESS from sending data, compromising the devices and jamming a communication network. A distributed secure control framework is presented, where an ACK-based attack detection strategy and a communication recovery mechanism are introduced to mitigate the impact of DoS attacks by repairing the paralyzed topology graphs caused by DoS attacks back into the initial connected graph. Under this framework, a distributed resilient finite-time secondary control scheme is proposed such that frequency regulation, active power sharing and energy level balancing of BESSs can be achieved simultaneously in a finite time, meanwhile, operational constraints can be satisfied at any control transient time. Moreover, based on theoretical analysis, the impact of the duration time of DoS attacks on the convergence time of the control algorithm can be explicitly revealed. Finally, validity and effectiveness of the proposed control scheme are demonstrated by case studies on a modified IEEE 57-bus testing system.

6.1 Problem Formulation

In this section, the BESS model is first presented, and then the control problem to be addressed in this chapter is formulated.

L. Ding et al., *Distributed Control and Optimization of Networked Microgrids*, Power Systems, https://doi.org/10.1007/978-3-030-95029-3_6

6.1.1 Control Structures for BESSs

Consider an AC microgrid incorporating N BESSs, where each BESS can communicate with its neighbors via a communication network. The communication topology is described as an undirected graph $\mathscr{G} = (\mathscr{V}, \mathscr{E})$. In the AC micgrogrid, each unit includes a BESS, a voltage source converter, an LC filter and an output connector. The converter is used to convert DC signals to AC ones and then transmit them into the grid. Figure 6.1 shows a basic hierarchical control structure for a BESS in the AC microgrid. According to [1], a simplified BESS model is established as

$$\dot{E}_i = -\frac{\eta_i}{3600} P_i \tag{6.1}$$

where E_i and P_i are the energy level and the active power of battery i, respectively; η_i is the heterogeneous parameter of battery i associated with its size. In practice, such battery parameters can be well estimated [2].

As shown in Fig. 6.1, the control structure of each battery includes primary control and secondary control. Generally, the aim of the primary control is to ensure frequency synchronization and voltage stability as well as active/reactive power sharing. In doing so, a conventional primary droop control (1.1) is commonly used [3].

It is well acknowledged that the decentralized droop control results in the deviations of frequency and voltage from their reference values. To compensate for the

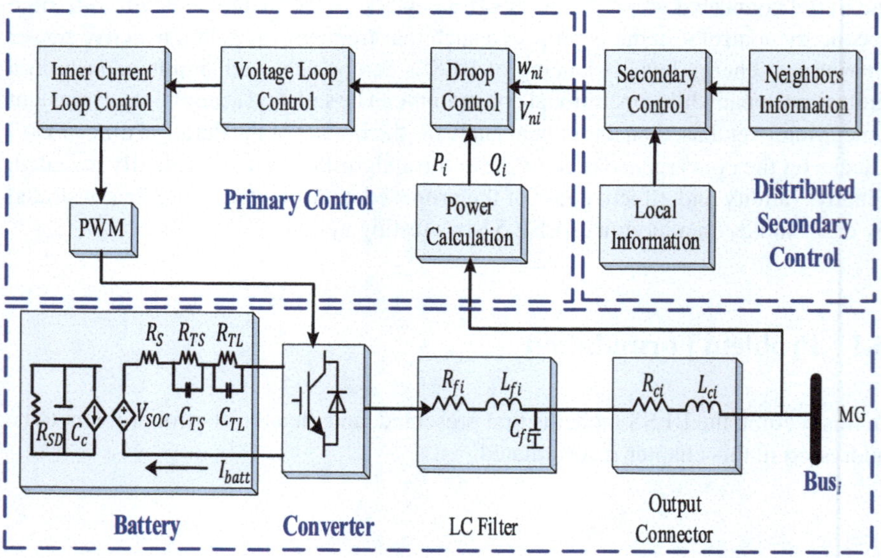

Fig. 6.1 Hierarchical control structure of a BESS

deviations and realize accurate frequency and voltage regulation, a secondary control scheme is introduced. Taking the derivative of (1.1a) yields

$$\dot{\omega}_{ni} = \dot{\omega}_i + m_i \dot{P}_i = u_i^{\omega} + u_i^{P} \tag{6.2}$$

where u_i^{ω} is the auxiliary control input of $\dot{\omega}_i$; u_i^{P} is the auxiliary control input of $m_i \dot{P}_i$. Then, the nominal set point is determined by

$$\omega_{ni} = \int (u_i^{\omega} + u_i^{P}) ds \tag{6.3}$$

Besides, in order to make full use of capacity of distributed BESSs, their energy levels should be dispatched to loads in a fair and reasonable way. One feasible strategy is to synchronize all the rated energy levels of batteries. Based on the consideration of frequency regulation, energy level synchronization, and power sharing, the combined dynamics incorporating the dynamics of BESSs (6.1) and the secondary control (6.2) can be given by [1]

$$\dot{E}_i = -\frac{\eta_i}{3600} P_i + u_i^{E} \tag{6.4a}$$

$$m_i \dot{P}_i = u_i^{P} \tag{6.4b}$$

$$\dot{\omega}_i = u_i^{\omega} \tag{6.4c}$$

where u_i^{E} is the control input of energy level. According to the results in [1], it is proved that adding the control input u_i^{E} into the dynamics of energy level (6.4a) is beneficial to enhance the convergence rate.

6.1.2 Operational Constraints

In a microgrid, operational capacity of each unit is usually limited, which should be incorporated to design control algorithms for the microgrid. In this chapter, the following physical constraints are included

$$\begin{cases} E_i^{min} \leq E_i(t + \Delta t) \leq E_i(t) \leq E_i^{max} \text{ if discharging} \\ E_i^{min} \leq E_i(t) \leq E_i(t + \Delta t) \leq E_i^{max} \text{ if charging} \end{cases} \tag{6.5a}$$

$$P_i^{min} \leq P_i(t) \leq P_i^{max} \tag{6.5b}$$

$$\omega_i^{min} \leq \omega_i(t) \leq \omega_i^{max} \tag{6.5c}$$

where E_i^{min} and E_i^{max} are the minimum and maximum limits of energy level for battery i, respectively; P_i^{min} and P_i^{max} are the minimum and maximum limits of active power output for battery i, respectively; ω_i^{min} and ω_i^{max} are the minimum and maximum limits of frequency for battery i, respectively.

According to [4], it is well acknowledged that a battery's lifespan is highly dependent upon its operation mechanism and frequent charge-discharge cycling might cause an extra cost for depreciation of BESSs, especially when fast regulation is implemented. Note that, when the constraint (6.5a) is satisfied, the energy levels of BESSs will be decreasing (or increasing) strictly in a discharging (or charging) mode, which can effectively avoid frequent switches between discharging and charging modes in the process of control implementation. Thus, from the perspective of battery's lifespan, it is practical and reasonable to take into account the constraint (6.5a) when designing control algorithms. Obviously, (6.5) are the hard constraints on energy levels, active power and frequency of batteries, providing their allowable operational ranges. If the hard constraints are violated at any transient period, it may result in instability of microgrids. Therefore, different from the existing literature [1, 5–8], in this chapter, we will design a distributed secondary controller ensuring that control operation of batteries is under such the constraints (6.5) at any times.

6.1.3 Control Objectives

Based on the discussion above, the *objective* of the chapter is that, under the constraints (6.5), designing distributed secondary controllers u_i^E, u_i^P and u_i^ω in (6.4) to achieve:

- *Frequency regulation:* to regulate the frequency of each BESS to the reference point in a finite time, namely,

$$\lim_{t \to t_\omega^*} \left| \omega_i(t) - \omega^{ref} \right| = 0, \forall i \tag{6.6}$$

$$\omega_i(t) = \omega^{ref}, \forall t \geq t_\omega^*, \tag{6.7}$$

  where ω^{ref} is the reference frequency of microgrids and t_ω^* is the settling time for frequency regulation.
- *Active power sharing*: to ensure proportional active power sharing among BESSs, namely,

$$\lim_{t \to t_p^*} \left| m_i P_i(t) - m_j P_j(t) \right| = 0; \;\; \forall i \neq j \tag{6.8}$$

$$m_i P_i(t) = m_j P_j(t); \;\; \forall t \geq t_p^* \tag{6.9}$$

  where t_p^* is the settling time for active power sharing;
- *Energy level balancing*: To synchronize the rated energy levels of BESSs, namely,

$$\lim_{t \to t_E^*} \left| \frac{E_i(t)}{\Delta E_i} - \frac{E_j(t)}{\Delta E_j} \right| = 0; \quad \forall i \neq j \tag{6.10}$$

$$\frac{E_i(t)}{\Delta E_i} = \frac{E_j(t)}{\Delta E_j}; \quad \forall t \geq t_E^* \tag{6.11}$$

where $\Delta E_i \triangleq E_i(0) - E_i^{min}$ for discharging or $\Delta E_i \triangleq E_i^{max} - E_i(0)$ for charging with initial energy level $E_i(0)$, and t_E^* is the settling time for energy level balancing.

Remark 6.1 Different from [6, 8], the issue of energy level balancing considered in this chapter is involved with the remaining or demanding energy of BESSs, namely, the energy levels of BESSs are synchronized proportional to their remaining or demanding energy. As a result, the strategy designed in this chapter can avoid any of BESSs to run out of energy or get full of energy in an earlier time.

6.2 Distributed Secondary Control Schemes Under DoS Attacks

In this section, distributed resilient secondary controllers are designed to achieve frequency regulation, active powering sharing and energy balancing under DoS cyber attacks. First, a description of DoS cyber attack is presented. Then, theoretical analysis is conducted to prove the convergence of the proposed control algorithms.

6.2.1 DoS Attack Modeling

Generally, DoS attacks refer to a class of attacks which make certain or all components of a control system inaccessible. The adversaries attempt to jam the communication channels, compromise the devices, block data sending and distort the routing protocols by imposing the DoS attacks on both the system and communication network. As a result, the occurrence of DoS attacks may seriously affect both system measurement and data transmission, causing loss of data and failure of control actions [9]. According to different attacked targets, DoS attacks on a networked system can be roughly categorized into: (1) node-based DoS attacks, which aim to prevent a node from sending its measurement data to any of its neighbors; (2) link-based DoS attacks, which aim to remove or damage the communication link between neighboring nodes. Note that, under the node-based attack, the connectivity of communication topology must be paralyzed since some or all of nodes are completely isolated from the communication network. Unfortunately, it is devastating for distributed control algorithms requiring necessary support of connected communication topology, probably resulting in degrading or even destroying the convergence of control algorithms. While, under the link-based attack where some communication links among nodes are removed, it is possible that the communication topology is kept either connected

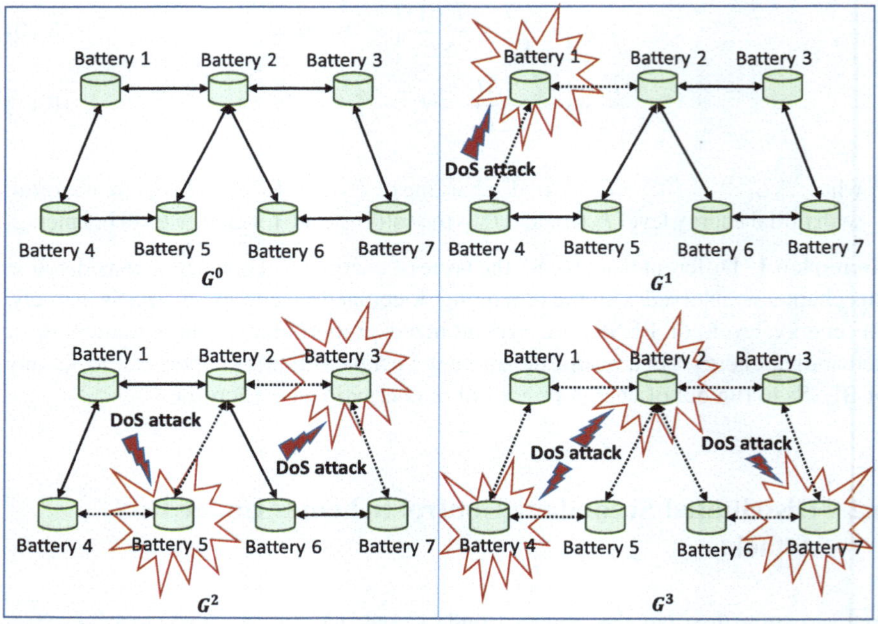

Fig. 6.2 Communication topologies under DoS attacks

or unconnected. It is clear that the former case only slows down the convergence of distributed control algorithms, while the latter case is in fact regarded as a case of the node-based attack. In contrast to the link-based attack, the node-based attack may probably result in relatively more serious degradation or even damage of control performance. Thus, in this chapter, we focus more on the case of node-based DoS attacks, where each battery is suffering from DoS attacks such that its measurement is blocked to be transmitted to all of its neighbors through a communication network. As a result, the battery under such DoS attacks gets isolated in communicating with others. It should be mentioned that the control framework designed later can also accommodate link-based DoS attacks.

(1) Impacts of DoS attacks: Obviously, the DoS attacks considered in the chapter will break communication connection among neighbors, leading to paralysis of graph connectivity of communication topology. To explain this more clearly, Figure 6.2 illustrates the impacts of DoS attacks on changes of communication topologies. It is emphasized that, graph G^0 is the original communication topology among batteries when attacks are absent, while some batteries in graphs G^1, G^2 and G^3 are under DoS attacks. For example, in graph G^1, the adversary launches a DoS attack on Battery 1 such that its measurement data cannot be delivered to Battery 2 and Battery 4, or vice versa. In this case, the Laplacian matrix of communication topology becomes

$$L = \begin{bmatrix} 0 & 0 & 0 & 0 & 0 & 0 & 0 \\ 0 & 3 & -1 & 0 & -1 & -1 & 0 \\ 0 & -1 & 2 & 0 & 0 & 0 & -1 \\ 0 & 0 & 0 & 1 & -1 & 0 & 0 \\ 0 & -1 & 0 & -1 & 2 & 0 & 0 \\ 0 & -1 & 0 & 0 & 0 & 2 & -1 \\ 0 & 0 & -1 & 0 & 0 & -1 & 2 \end{bmatrix}$$

It is noted that the all the elements in both the first row and column of L are zero, which implies that Battery 1 has no any communication with others. According to [10], it is theoretically not possible for Battery 1 to achieve cooperative control. More specially, all the communication links in graph G^3 are broken although only three batteries are subject to the DoS attacks. Therefore, the presence of DoS attacks significantly paralyzes the connectivity of communication topology, leading to failure of control algorithms in terms of frequency regulation, active power sharing and energy balancing.

(2) Detection of DoS attacks: Due to the negative impact of DoS attacks on control performance, how to efficiently detect DoS attacks becomes critical in the microgrid, especially when attackers can continuously put the jamming on nodes. In order to mitigate the impact of DoS attacks, an ACK-based attack detection approach is introduced in this chapter, which is simple to implement in practice. More specifically, Battery i sends its data information at time t to its neighbors j, and then its neighbors that have access to the data information will immediately send an acknowledgment message back to Battery i. Here, it is assumed that every ACK-message can be successfully transmitted through secure channels with different protocols from communication channels among batteries. Such implementation is practical and commonly used in the area of cybersecurity framework for electrical grids [11, 12]. Thus, if Battery i does not receive the ACK-message in a period $[t, t+\tau^{ack}]$ with τ^{ack} being the maximum time for receiving the ACK-message, then it implies that Battery i is suffering from the DoS attacks with time t being time instant of the kth DoS attack, denoted by t^a_{ik}. When there are multiple DoS attacks on different nodes before the network recovery, one has the kth attack instant $t^a_k = \inf\{t|t^a_{ik}, i \in \mathcal{N}\}$.

(3) Communication recovery mechanism: To ensure security and reliability of network communication, a smart communication management center (SCMC) is employed to perform the monitoring of communication environment and decision-making of network repairment. Similarly, with the cybersecurity framework [11, 12], a reliable and secure channel is employed to transmit the attack detection report between each battery and the SCMC. When detecting the event of attacks, batteries will send a report massage to the SCMC. Then, the SCMC will promptly activate operation of network repairment to reinitialize or recover the entire communication topology of batteries. It should be mentioned that there have been many well-studied approaches to reinitialization or recovery of broken communication topology in the existing literature such as [13–15]. The details on these communication recovery mechanisms can be referred to [13–15] for interested readers, which are beyond the scope of this chapter. In practice, the processing of communication network

repairement will take a finite period of time, which closely depends on the numbers of attacked nodes and broken communication links.

(4) Model of DoS attacks: Based on the observation above, the DoS attacks can be modeled by a set of communication topologies which are swiching between the connectivity-paralyzed graphs and the initial connected graph. To be more specific, the communication topology without any attacks and after repairment is set by G^0 with its weights a_{ij}^0 and its Laplacian matrix L^0. Here, it is assumed that, the kth attacks happening to the microgrid can be detected at the time sequence t_k^a $(k = 1, 2, \ldots)$, and the time when the communication topology is recovered from the kth attack is denoted by t_k $(k = 0, 1, \ldots)$ with $t_0 = 0$ being the initial time. Thus, it is easy to know that the duration intervals without and with attacks are $[t_{k-1}, t_k^a)$ and $[t_k^a, t_k)$, respectively. In fact, the attacked time length of $[t_k^a, t_k]$ is the mount of attack detection time, message transmission time and network recovery time. Let δ_{ij} represent the status of communication links at node i, where $\delta_{ij}(t) = 0$ for any j if node i is attacked during $t \in [t_k^a, t_k)$, otherwise, $\delta_{ij}(t) = 1$. Then, the communication topology under attack can be described as a time-varying graph $G^{\delta(t)}$ with its weights $a_{ij}^{\delta_{ij}(t)} = \delta_{ij}(t) a_{ij}^0$ and its Laplacian matrix $L^{\delta(t)}$.

6.2.2 Distributed Resilient Finite-Time Secondary Controller

To facilitate the design of distributed secondary controller, the following assumption is needed.

Assumption 6.2.1 The communication topology graph $\mathscr{G}$ is undirected and connected when DoS attacks are absent. □

To simplify presentation, let $\hat{P}_i(t) = m_i P_i(t)$ and $\hat{E}_i(t) = E_i(t)/\Delta E_i$, and we drop t from the states. Then, regarding the effects of DoS attacks, the following distributed secondary controllers are proposed: No attacks, i.e. $t \in [t_{k-1}, t_k^a)$

$$u_i^E(t) = \Pi^{\beta_i}\left(-k_1 \Delta E_i \sum_{j \in N_i} a_{ij}^0 \left(\hat{E}_i(t) - \hat{E}_j(t)\right)^{\frac{a}{b}}\right) \tag{6.12a}$$

$$u_i^P(t) = \mathrm{Proj}_{P_i}\left(-k_2 \sum_{j \in N_i} a_{ij}^0 \left(\hat{P}_i(t) - \hat{P}_j(t)\right)^{\frac{a}{b}}\right) \tag{6.12b}$$

$$\begin{aligned} u_i^{\omega}(t) = \mathrm{Proj}_{\omega_i}\Bigg(&-k_3 \sum_{j \in N_i} a_{ij}^0 \left(\omega_i(t) - \omega_j(t)\right)^{\frac{a}{b}} \\ &- k_3 g_i \left(\omega_i(t) - \omega^{ref}\right)^{\frac{a}{b}}\Bigg) \end{aligned} \tag{6.12c}$$

Under attacks, i.e. $t \in [t_k^a, t_k)$

$$u_i^E(t) = \Pi^{\beta_i}\left(-k_1 \Delta E_i \sum_{j \in N_i} a_{ij}^{\delta_{ij}(t)} \left(\hat{E}_i(t) - \hat{E}_j(t)\right)^{\frac{a}{b}}\right) \tag{6.13a}$$

$$u_i^P(t) = \text{Proj}_{P_i}\left(-k_2 \sum_{j \in N_i} a_{ij}^{\delta_{ij}(t)} \left(\hat{P}_i(t) - \hat{P}_j(t)\right)^{\frac{a}{b}}\right) \tag{6.13b}$$

$$u_i^w(t) = \text{Proj}_{w_i}\left(-k_3 \sum_{j \in N_i} a_{ij}^{\delta_{ij}(t)} \left(\omega_i(t) - \omega_j(t)\right)^{\frac{a}{b}} - k_3 g_i \left(\omega_i(t) - \omega^{ref}\right)^{\frac{a}{b}}\right) \tag{6.13c}$$

where $\beta_i = \frac{\eta_i \hat{P}_i}{3600 m_i}$; a and b are positive odd integers with $a < b$; $k_p > 0 (p = 1 - 3)$ are controller gains; g_i is the pinning gain for battery i, where $g_i = 1$ if battery i can have access to the reference frequency ω^{ref}; $g_i = 0$, otherwise. It is assumed in this chapter that at least one battery can have access to ω^{ref}. The projection operator Π^{β} is to ensure the constraint (6.5a) can be satisfied, which is defined by

$$\Pi^{\beta}(v(x)) \triangleq \begin{cases} \beta & \text{if } x = x^{min} \text{ or } x = x^{max} \\ & \text{or } v > \beta \ (discharging) \\ & \text{or } v < \beta \ (charging) \\ v & \text{otherwise} \end{cases} \tag{6.14}$$

The projection operator Proj_x is used to constrain the magnitude of frequency and active power in control actions

$$\text{Proj}_x(v(x)) \triangleq \begin{cases} 0 & \text{if } v > 0 \,\&\, x = x^{max} \\ & \text{or } v < 0 \,\&\, x = x^{min} \\ v & \text{otherwise} \end{cases} \tag{6.15}$$

where x^{max} and x^{min} are the upper and lower bounds of x, respectively. According to the definition of $\text{Proj}_x(\cdot)$, it is easy to obtain its following characteristics

$$(x - x^*)\text{Proj}_x^{\alpha}(v(x)) \le (x - x^*)v(x) \tag{6.16}$$

for any $x^* \in [x^{min}, x^{max}]$. It should be pointed out that, introducing the projection operator $\text{Proj}(\cdot)$ in controllers (6.12) and (6.13) is to ensure that the hard constraints (6.5) can be satisfied at any transient time.

Remark 6.2 Different from the control algorithms in [1, 5, 6, 8, 16], the control algorithms (6.12) and (6.13) in this chapter include some new properties: (i) the projection operators are introduced to ensure the constraints (6.5) can be satisfied at any control transient time; and (ii) the communication topologies are switching among the initial connected graph and a set of connectivity-paralyzed graphs caused by DoS attacks. Consequently, they lead to some difficulties in theoretical analysis and the estimate on the settling time of the algorithms, which will be dealt with in the next subsection.

6.2.3 Stability Analysis

In order to conduct convergence analysis of controllers (6.12) and (6.13), we define the error variables

$$e_{pi}(t) = \tilde{P}(t) - \frac{1}{N}\sum_{i=1}^{N}\tilde{P}(t)$$

$$e_{Ei}(t) = \hat{E}_i(t) - \frac{1}{N}\sum_{j=1}^{N}\hat{E}_j(t)$$

$$e_{\omega i}(t) = \omega_i(t) - \omega^{ref}$$

with their compact form

$$e_p(t) = [e_{p1}(t), e_{p2}(t), \ldots, e_{pN}(t)]^T,$$

$$e_E(t) = [e_{E1}(t), e_{E2}(t), \ldots, e_{EN}(t)]^T,$$

$$e_\omega(t) = [e_{\omega 1}(t), e_{\omega 2}(t), \ldots, e_{\omega N}(t)]^T.$$

Let the adjacency weights $\tilde{a}_{ij}^0 = (a_{ij}^0)^{\frac{2b}{a+b}}$ and its corresponding Laplacian matrix be $\tilde{L}^0$. Denote by $\lambda_i(\cdot)$ the ith smallest eigenvalue of a matrix.

Now, we establish the following theorem.

Theorem 6.1 *Under Assumption 6.2.1, for given scalars $k_l > 0$ $(l = 1, 2, 3)$ and parameters*

$$\frac{\eta_i}{m_i \Delta E_i} = \frac{\eta_j}{m_j \Delta E_j}, \quad \forall i, j \in \mathcal{V},$$

using the distributed secondary controllers (6.12) and (6.13), the BESSs in a micro-grid can simultaneously achieve energy level balancing, active power sharing and frequency regulation in a finite time. Moreover, the upper bound of total settling time is estimated by

$$T^* = \max\{T_p, T_E, T_\omega\} + T_a \tag{6.17}$$

where T_a is the total duration time of attacks and

$$T_p = \frac{b\|e_p(0)\|^{\frac{b-a}{b}}}{(b-a)k_2 2^{\frac{a}{b}} \lambda_2^{\frac{a+b}{2b}}(\tilde{L}^0)}, \tag{6.18}$$

$$T_E = \frac{b\|e_E(0)\|^{\frac{b-a}{b}}}{(b-a)k_1 2^{\frac{a}{b}} \lambda_2^{\frac{a+b}{2b}}(\tilde{L}^0)}, \tag{6.19}$$

$$T_w = \frac{b\|e_w(0)\|^{\frac{b-a}{b}}}{(b-a)k_3 2^{\frac{a}{b}} \lambda_1^{\frac{a+b}{2b}}(\tilde{M}^0)} \tag{6.20}$$

with $\tilde{M}^0 = \tilde{L}^0 + \text{diag}\{g_i\}_N$.

Proof Based on the definition of $e_{pi}(t)$ and the controllers (6.12b) and (6.13b), one has

$$\dot{e}_{pi}(t) = \dot{\hat{P}}(t) - \frac{1}{N}\sum_{i=1}^{N} \dot{\hat{P}}(t)$$
$$= \begin{cases} \text{Proj}\left(-k_2 \sum_{j\in N_i} a_{ij}^0 (e_{pi}(t) - e_{pj}(t))^{\frac{a}{b}}\right) \text{ for } t \in [t_{k-1}, t_k^a) \\ \text{Proj}\left(-k_2 \sum_{j\in N_i} a_{ij}^{\delta_{ij}(t)} (e_{pi}(t) - e_{pj}(t))^{\frac{a}{b}}\right) \text{ for } t \in [t_k^a, t_k) \end{cases}$$

Choose a Lyapunov function as

$$V_p(t) = \frac{1}{2}e_p^T(t)e_p(t) = \frac{1}{2}\sum_{i=1}^{N} e_{pi}^2(t),$$

and then taking a time derivative of $V_p(t)$ yields:

Case I: for $t \in [t_{k-1}, t_k^a)$

$$\begin{aligned}\dot{V}_p(t) &= \sum_{i=1}^{N} e_{pi}(t)\dot{e}_{pi}(t) \\ &= \sum_{i=1}^{N} e_{pi}(t)\text{Proj}\Big(-k_2 \sum_{j\in N_i} a_{ij}^0 \big(e_{pi}(t) - e_{pj}(t)\big)^{\frac{a}{b}}\Big) \\ &\le -k_2 \sum_{i=1}^{N} e_{pi}(t)\Big(\sum_{j\in N_i} a_{ij}^0 \big(e_{pi}(t) - e_{pj}(t)\big)^{\frac{a}{b}}\Big) \\ &= -\frac{k_2}{2}\sum_{i=1}^{N}\sum_{j\in N_i}\Big(\tilde{a}_{ij}^0 \big(e_{pi}(t) - e_{pj}(t)\big)^2\Big)^{\frac{a+b}{2b}} \\ &\le -\frac{k_2}{2}\Big(\sum_{i=1}^{N}\sum_{j\in N_i}\tilde{a}_{ij}^0 \big(e_{pi}(t) - e_{pj}(t)\big)^2\Big)^{\frac{a+b}{2b}} \end{aligned} \tag{6.21}$$

where the first inequality is based on the projection characteristic (6.16).

Since Assumption 6.2.1 holds, one knows that $\tilde{L}^0$ has a simple eigenvalue of zero. Then, we have

$$\sum_{i=1}^{N}\sum_{j\in N_i}\tilde{a}_{ij}^0 \big(e_{pi}(t) - e_{pj}(t)\big)^2 = 2e_p^T(t)\tilde{L}^0 e_p(t) \ge 2\lambda_2(\tilde{L}^0)e_p^T(t)e_p(t).$$

Thus, from (6.21), one can obtain

$$\begin{aligned}\dot{V}_p(t) &\le -\frac{k_2}{2}\Big(2\lambda_2(\tilde{L}^0)e_p^T(t)e_p(t)\Big)^{\frac{a+b}{2b}} \\ &= -\tilde{c}_1 V_p^{\frac{a+b}{2b}}(t) \end{aligned} \tag{6.22}$$

where $\tilde{c}_1 = k_2 2^{\frac{a}{b}} \lambda_2^{\frac{a+b}{2b}}(\tilde{L}^0)$. Based on (6.22), we have

$$\Big(V_p^{\frac{b-a}{2b}}(t)\Big)' = \frac{b-a}{2b} V_p^{-\frac{a+b}{2b}}(t)\dot{V}_p(t) \le -c_1 \tag{6.23}$$

where $c_1 = \frac{\tilde{c}_1(b-a)}{2b}$.

Integrating both sides of (6.23) on the interval $t \in [t_{k-1}, t_k^a)$ leads to

$$\int_{t_{k-1}}^{t} \Big(V_p^{\frac{b-a}{2b}}(t)\Big)' dt \le -\int_{t_{k-1}}^{t} c_1 dt,$$

which implies that, for $t \in [t_{k-1}, t_k^a)$,

$$V_p^{\frac{b-a}{2b}}(t) \leq V_p^{\frac{b-a}{2b}}(t_{k-1}) - c_1(t - t_{k-1}) \tag{6.24}$$

Case II: for $t \in [t_k^a, t_k)$

Let $\tilde{L}^{\delta(t)}$ denote the Laplacian matrix with its adjacency weights $\tilde{a}_{ij}^0\delta_{ij}$. Since at least one battery is attacked at the time t_k^a, the matrix $\tilde{L}^{\delta(t)}$ has more than two eigenvalues of 0, which implies that $\lambda_2(\tilde{L}^{\delta(t)}) = 0$.

Following the similar lines of (6.21) and (6.22), one can get

$$\dot{V}_p(t) \leq -\frac{k_2}{2}\left(2\lambda_2(\tilde{L}^{\delta(t)})e_p^T(t)e_p(t)\right)^{\frac{a+b}{2b}} = 0 \tag{6.25}$$

which implies that,

$$V_p^{\frac{b-a}{2b}}(t) \leq V_p^{\frac{b-a}{2b}}(t_k^a) \text{ for } t \in [t_k^a, t_k) \tag{6.26}$$

Based on Case I and Case II, we have:

(1) if $t \in [t_{k-1}, t_k^a)$,

$$\begin{aligned} V_p^{\frac{b-a}{2b}}(t) &\leq V_p^{\frac{b-a}{2b}}(t_{k-1}) - c_1(t - t_{k-1}) \\ &\leq V_p^{\frac{b-a}{2b}}(t_{k-1}^a) - c_1(t - t_{k-1}) \\ &\leq V_p^{\frac{b-a}{2b}}(t_{k-2}) - c_1(t_{k-1}^a - t_{k-2}) - c_1(t - t_{k-1}) \\ &= V_p^{\frac{b-a}{2b}}(t_{k-2}) + c_1(t_{k-1} - t_{k-1}^a) - c_1(t - t_{k-2}) \\ &\leq \cdots \\ &\leq V_p^{\frac{b-a}{2b}}(0) + c_1T_a - c_1t \end{aligned} \tag{6.27}$$

(2) if $t \in [t_k^a, t_k)$

$$\begin{aligned} V_p^{\frac{b-a}{2b}}(t) &\leq V_p^{\frac{b-a}{2b}}(t_k^a) \\ &\leq V_p^{\frac{b-a}{2b}}(t_{k-1}) - c_1(t_k^a - t_{k-1}) \\ &= V_p^{\frac{b-a}{2b}}(t_{k-1}) + c_1(t - t_k^a) - c_1(t - t_{k-1}) \\ &\leq \cdots \\ &\leq V_p^{\frac{b-a}{2b}}(0) + c_1T_a - c_1t \end{aligned} \tag{6.28}$$

Obviously, when $t \geq t_p^* = T_p + T_a$, $V_p(t)$ will converge to 0, which implies that $e_p(t) \to 0$ after t_p^*. Therefore, it is concluded that the active power sharing can be achieved in a prescribed time t_p^*, namely,

$$\hat{P}_i = \hat{P}_j = \frac{1}{N}\sum_{i=1}^{N} \hat{P}(0), \forall i, j \in \mathscr{V}$$

for $t \geq t_p^*$.

In what follows, we will prove the finite-time convergence of the energy balancing control algorithms (6.12a) and (6.13a) for BESSs. By solving (6.4a), one can obtain

$$\hat{E}_i(t) = \bar{E}_i(t) - \frac{\eta_i}{3600 m_i \Delta E_i}\int_0^t \hat{P}_i(s)ds \tag{6.29}$$

where $\bar{E}_i(t)$ is the solution of

$$\dot{\bar{E}}_i(t) = u_i^E / \Delta E_i$$

with the control algorithms (6.12a) and (6.13a).

Let

$$V_E = \frac{1}{2}\sum_{i=1}^{N} e_{Ei}^2(t).$$

Recalling the proof of active power sharing above, it is known that

$$\bar{E}_i(t) = \bar{E}_j(t) = \frac{1}{N}\sum_{i=1}^{N} \hat{E}(0), \forall i, j \in \mathscr{V}$$

for $t \geq t_E^{'} = T_E + T_a$.

Given parameters $\frac{\eta_i}{m_i \Delta E_i} = \frac{\eta_j}{m_j \Delta E_j} = \kappa, \forall i, j \in \mathscr{V}$, it is concluded that, for $t \geq t_E^* = \max\{t_p^*, t_E^{'}\}$

$$\hat{E}_i(t) = \frac{1}{N}\sum_{i=1}^{N} \hat{E}(0) - \frac{\kappa \sum_{i=1}^{N} \hat{P}(0)}{N 3600} \cdot t, \ \ \forall i \in \mathscr{V} \tag{6.30}$$

which means that the energy level balancing can be achieved within a prescribed time t_E^*.

Next, we will prove that the frequencies of BESSs can be regulated to the reference value in a finite time. Define a Lyapunov function candidate $V_\omega = \frac{1}{2}\sum_{i=1}^{N} e_{\omega i}^2$. Following the similar lines to the proofs above, one obtains that $\omega_i = \omega^{ref}, \forall i \in \mathscr{V}$ for $t \geq t_\omega^* = T_\omega + T_a$, which means that frequency regulation of BESSs can be achieved within a prescribed time t_ω^*.

Finally, according to Theorem 4.2 in [17], it is straightforward to conclude that the projection operator can ensure that all the trajectories of $P_i(t)$, $E_i(t)$ and $\omega_i(t)$ stay within the hard constraints (6.5) if their initial conditions satisfy (6.5).

From the observation above, it is concluded that the estimation of total convergence time is $T^* = \max\{t_p^*, t_E^*, t_\omega^*\}$, which is rewritten as (6.17). The proof of Theorem 6.1 is complete.

Remark 6.3 It should be pointed out that, Theorem 6.1 explicitly reveals the relationship between the duration time T_a of DoS attacks and the convergence time T^* of the proposed control algorithms (6.12) and (6.13). More specifically, given the control gains, system parameters and initial conditions, the longer duration time of DoS attacks will result in the slower convergence rate. If an expected convergence time T_{exp}^* is preset, then the attack duration time T_a is required to be less than $T_{exp}^* - \max\{T_p, T_E, T_w\}$, which is meaningful for practical implementation.

6.3 Case Studies

In this section, case studies with a modified IEEE 57-bus testing system are conducted to show effectiveness and efficiency of the proposed control scheme. As shown in Fig. 6.3, the testing system includes 7 lithiumion BESSs and 42 loads which are connected to the micrigrid. The BESSs have heterogeneous capacity, where the capacities of Batteries 1–3, Batteries 4–6 and Battery 7 are 200k Wh, 150 kWh and 250 kWh, respectively. BESSs are connected to the main grid through the Point of Common Coupling (PCC), which is used to determine the operation of the microgrid. The desired output frequency and voltage of BESSs are $f^{ref} = 50$ Hz ($\omega^{ref} = 2\pi f^{ref} \approx 314.15$ rad/s) and $V^{ref} = 20$ kV, respectively. The upper and lower limits of output power generation at BESSs are

$$P^{max} = [50, 80, 45, 100, 60, 50, 70]^T\,(\text{kW}),$$
$$P^{min} = [10, 20, 15, 10, 15, 5, 7]^T\,(\text{kW}),$$

respectively. The range of frequency is given by $f_i \in [49, 51]$ (Hz). In order to protect the lifetime of batteries, they should work between 80% and 20% of total energy. Thus, it is easy to know that

$$E^{min} = [40, 40, 40, 30, 30, 30, 50]^T\,(\text{kWh}),$$
$$E^{max} = [160, 160, 160, 120, 120, 120, 200]^T\,(\text{kWh})$$

6.3.1 Performance Evaluation

We first verify the validity and effectiveness of the proposed distributed control scheme. It is assumed that all the BESSs of the microgrid operate in a discharging mode. Let the initial energy levels of batteries be $E(0) = [160, 150, 155, 105, 110, 100, 195]^T$, and the total loads be 310 kW. We choose the parameters

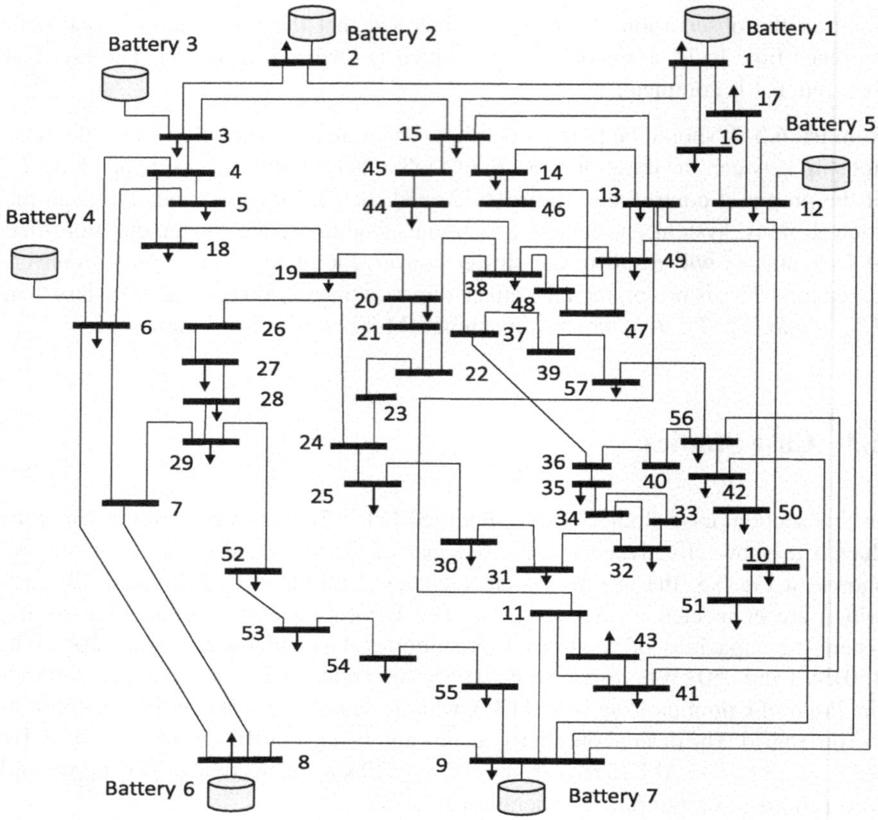

Fig. 6.3 A modified IEEE 57-bus testing system

$$k_1 = 0.25, \quad k_2 = 0.45, \quad k_3 = 1, \quad a = 3, \quad b = 7$$

$$\eta = [1, 1, 1, 0.75, 0.75, 0.75, 1.25]^T$$

It is assumed that BESSs can communicate with their neighbors via a communication network. In addition, it is assumed that battery 1 and battery 5 can have access to the reference frequency, i.e. $g_1 = g_5 = 1$. Suppose that the microgrid is subject to DoS attacks. When there is no any communication recovery mechanism (after the first attack in Fig. 6.4), the simulation results for all the BESSs are given in Fig. 6.5. Obviously, the control scheme fails to realize the objectives of frequency restoration, active power sharing and energy level balancing of batteries. Under the communication recovery mechanism, the switching of communication topology can be shown in Fig. 6.4. Using the distributed control schemes (6.12) and (6.13), the simulation results are shown in Fig. 6.6. Based on (6.18–6.20), we can calculate $T_p = 31.6317$ s, $T_E = 77.4694$ s, $T_\omega = 2.3871$ s, and then $T_a = 7$ s. It can be observed from Fig. 6.6 that, the frequency regulation, the active power sharing and the energy level balanc-

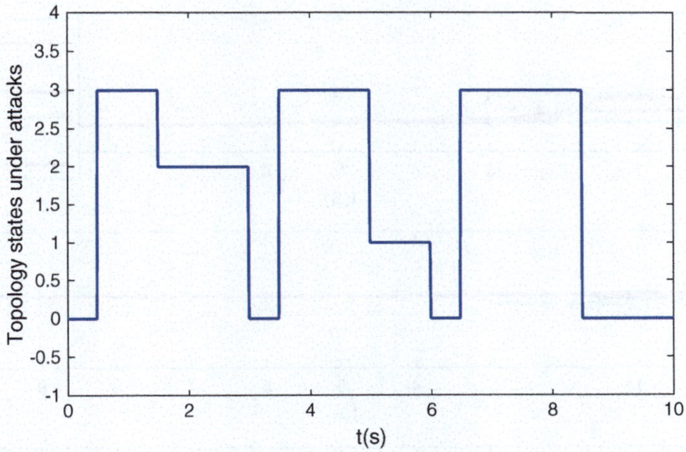

Fig. 6.4 Topology switching under DoS attacks: the states of 0, 1, 2, 3 on the y-axis correspond to the topologies G_i of Fig. 6.2, respectively

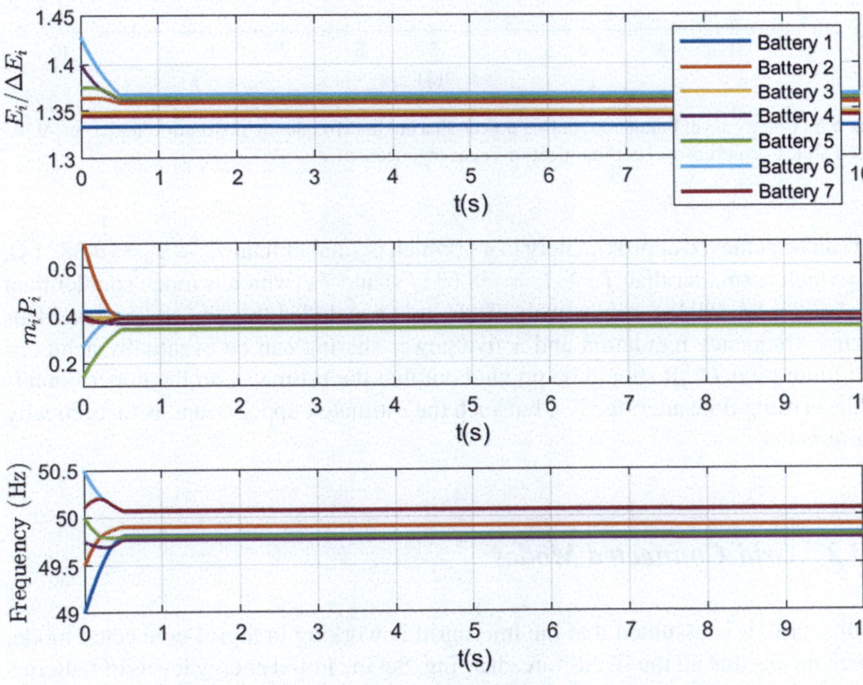

Fig. 6.5 Energy level balancing, active power sharing and frequency regulation without any attack detection and communication recovery mechanisms

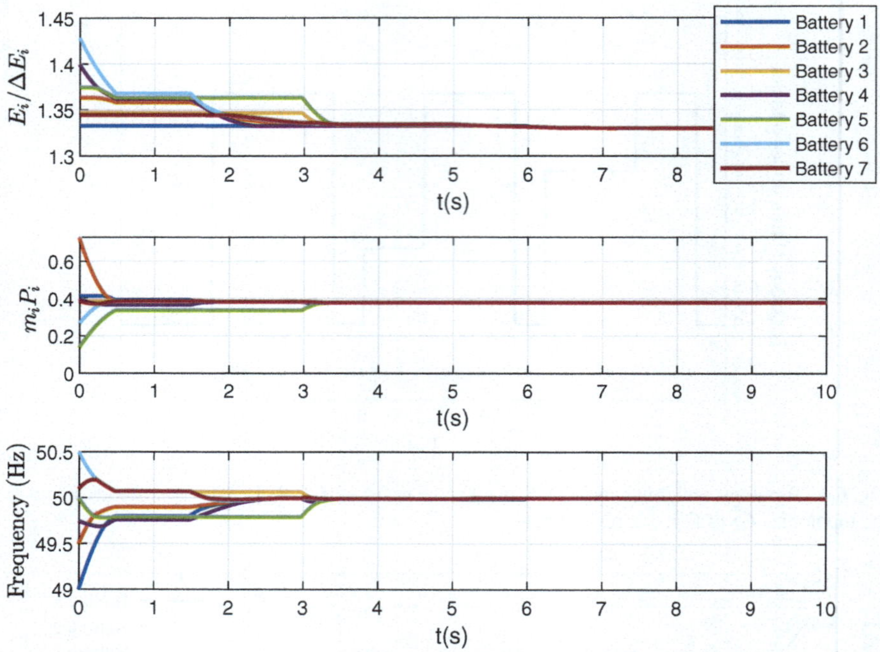

Fig. 6.6 Energy level balancing, active power sharing and frequency restoration under an ACK-based attack detection and communication recovery mechanisms

ing can be achieved approximately in 4 s (which is smaller than $T_\omega + T_a = 9.3871$ s), 4 s (which is smaller than $T_p + T_a = 38.6317$ s) and 7 s (which is much smaller than $T_E + T_a = 84.4694$ s), respectively. Thus, it is concluded that the energy level balancing, frequency regulation and active power sharing can be eventually achieved in a finite time T^*. It should be pointed out that the estimates on the upper bounds of the settling time are effective, but such the estimated upper bound is theoretically conservative.

6.3.2 Grid-Connected Modes

In this case, it is assumed that the microgrid is working in a grid-connected mode, which means that all the BESSs are charging. Set the initial energy levels of batteries $E(0) = [60, 50, 55, 45, 30, 40, 95]^T$. Let $k_1 = 1.25$ and all other parameters be same as the case of Sect. 6.3.1. Applying the control algorithms (6.12) and (6.13), we plot the evolution of energy levels, active power sharing and frequency regulation in Fig. 6.7. It is clear to see that the proposed control algorithms (6.12) and (6.13) are effective for both the grid-connected and islanded modes.

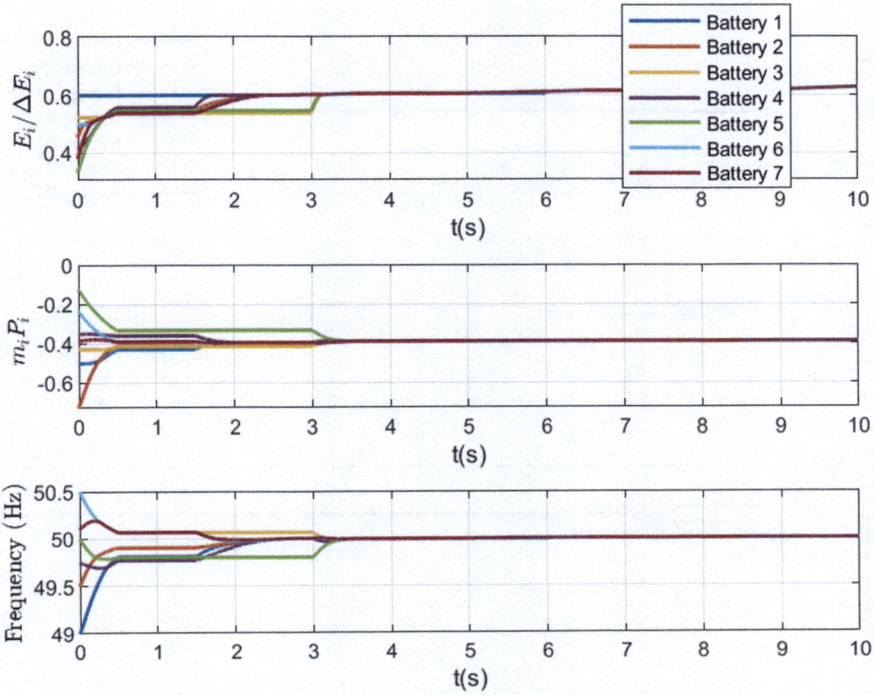

Fig. 6.7 Energy level balancing, active power sharing and frequency restoration in a grid-connected mode

6.3.3 Comparison Results

In this subsection, we will make a set of comparisons to show the efficiency of the proposed control schemes.

(1) Functions of projection operators: In order to guarantee the operational constraints (6.5) in control actions, we introduce the projection operators Π and Proj in the controllers (6.12) and (6.13). Figure 6.8 shows the simulation results without projection operators. From Fig. 6.8, it is clear to see that the energy levels of some batteries are increasing in the beginning period time, which is not reasonable in the discharging mode. Alternatively, the energy levels in Fig. 6.6 are non-increasing in the entire control process. In this sense, it is clarified that the projection operators are effective to ensure the operational constraints (6.5) are satisfied in any control transient instants.

(2) The effects of parameters a/b on convergence rates: In this case, comparison results of convergence rate using different parameters a and b are given. To show the comparison clearly, we set four cases with $a/b = 1$, $a/b = 3/5$, $a/b = 5/9$ and $a/b = 3/7$, respectively, and plot the convergence of energy levels in Fig. 6.9. It is observed from Fig. 6.9 that, as the value of a/b gets smaller, the convergence of

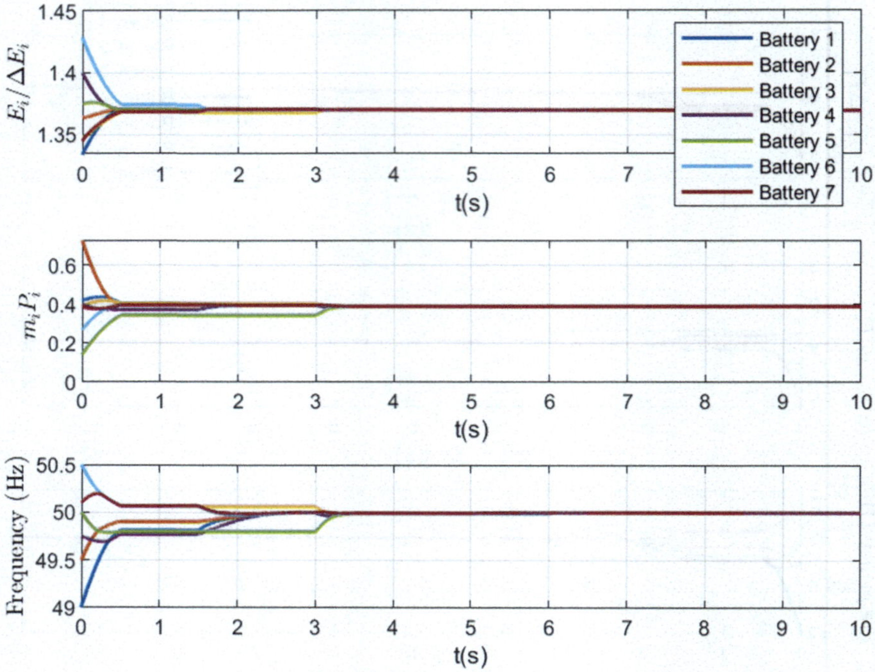

Fig. 6.8 Distributed control scheme without operational constraints

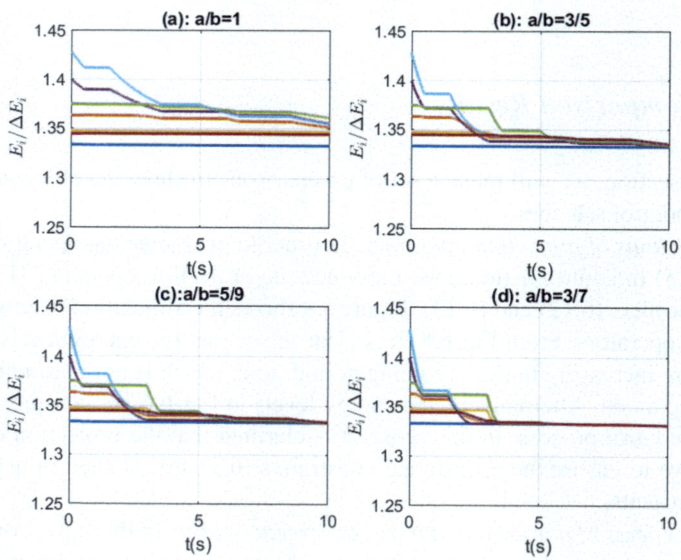

Fig. 6.9 Comparison of convergence rates using different a/b

BESS systems will become faster. When $a/b = 1$, the distributed control schemes (6.12) and (6.13) reduce to the similar form of [6], which implies that the proposed distributed control scheme has a faster convergence rate than that of [6].

6.4 Concluding Remarks

A distributed resilient secondary control scheme that takes the effects of DoS attacks into account has been presented to achieve frequency regulation, active power sharing and energy level balancing in a finite time. In the control design, a communication recovery mechanism has been used to repair the connectivity-paralyzed communication topology so as to ensure robustness of the proposed control scheme against DoS attacks. Theoretical analysis has been conducted to verify the convergence of the proposed control scheme. Finally, simulation case studies on a modified IEEE 57-bus testing system have been performed to demonstrate effectiveness of the proposed methods.

6.5 Notes

The results in this chapter are based mainly on [18]. In this chapter, a simplified battery model (6.1) is used, which has been widely adopted in existing literature [1, 5, 6, 8]. Moreover, it is shown in [1] that, by ignoring the fast dynamics of internal voltage and current controllers, the simplified BESS can be employed to model a large-scale power systems with multiple BESSs. It is true that, in more practical and accurate battery modeling, the battery model includes more complex dynamics such as nonlinearity and uncertainty, which may be caused by its internal inherent structure and external environment conditions. For such a complex battery model, a robust and more complicated controller will be required to ensure the performance of battery. Obviously, it is an interesting and challenging issue to extend the algorithms proposed in the chapter into the model with nonlinearity and uncertainty, which will be further take into account in our future work.

In addition, for analysis brevity, suppose that the transmission of all the state information such as frequencies, active power and energy levels of BESSs uses a common communication topology graph $\mathcal{G}$. In a practical microgrid, however, it is possible that the information is shared through different communication networks (denoted by $\mathcal{G}^{\omega}$, $\mathcal{G}^{E}$ and $\mathcal{G}^{p}$, respectively) like [19]. It should be emphasized that, the control method proposed in this chapter can be extended to accommodate the case of different topologies.

References

1. J. Khazaei, Z. Miao, Consensus control for energy storage systems. IEEE Trans. Smart Grid **9**(4), 3009–3017 (2018)
2. Q. Yu, R. Xiong, C. Lin, W. Shen, J. Deng, Lithium-ion battery parameters and state-of-charge joint estimation based on h-infinity and unscented kalman filters. IEEE Trans. Veh. Technol. **66**(10), 8693–8701 (2017)
3. F. Chen, M. Chen, Q. Li, K. Meng, Y. Zheng, J.M. Guerrero, D. Abbott, Cost-based droop schemes for economic dispatch in islanded microgrids. IEEE Trans. Smart Grid **8**(1), 63–74 (2017)
4. G. He, Q. Chen, C. Kang, P. Pinson, Q. Xia, Optimal bidding strategy of battery storage in power markets considering performance-based regulation and battery cycle life. IEEE Trans. Smart Grid **7**(5), 2359–2367 (2016)
5. D.H. Nguyen, J. Khazaei, Multiagent time-delayed fast consensus design for distributed battery energy storage systems. IEEE Trans. Sustain. Energy **9**(3), 1397–1406 (2018)
6. J. Khazaei, D.H. Nguyen, Multi-agent consensus design for heterogeneous energy storage devices with droop control in smart grids. IEEE Trans. Smart Grid **10**(2), 1395–1404 (2019)
7. T. Morstyn, B. Hredzak, V.G. Agelidis, Cooperative multi-agent control of heterogeneous storage devices distributed in a DC microgrid. IEEE Trans. Power Syst. **31**(4), 2974–2986 (2016)
8. J. Hu, A. Lanzon, Distributed finite-time consensus control for heterogeneous battery energy storage systems in droop-controlled microgrids. IEEE Trans. Smart Grid **10**(5), 4751–4761 (2019)
9. C. Peng, H. Sun, M. Yang, Y.-L. Wang, A survey on security communication and control for smart grids under malicious cyber attacks. IEEE Trans. Syst. Man Cybern. Syst. **49**(8), 1554–1569 (2019)
10. R. Olfati-Saber, R. Murray, Consensus problems in networks of agents with switching topology and time-delays. IEEE Trans. Autom. Control **49**(9), 1520–1533 (2004)
11. J. Qi, A. Hahn, X. Lu, J. Wang, C.-C. Liu, Cybersecurity for distributed energy resources and smart inverters. IET Cyber-Phys. Syst.: Theory Appl. **1**(1), 28–39 (2016)
12. T. Cruz, L. Rosa, J. Proença, L. Maglaras, M. Aubigny, L. Lev, J. Jiang, P. Simoes, A cybersecurity detection framework for supervisory control and data acquisition systems. IEEE Trans. Ind. Inform. **12**(6), 2236–2246 (2016)
13. A. Majdandzic, B. Podobnik, S.V. Buldyrev, D.Y. Kenett, S. Havlin, H.E. Stanley, Spontaneous recovery in dynamical networks. Nat. Phys. **10**(1), 34 (2014)
14. M.A. Di Muro, C.E. La Rocca, H. Stanley, S. Havlin, L.A. Braunstein, Recovery of interdependent networks. Sci. Rep. **6**, 22834 (2016)
15. Y. Shang, Localized recovery of complex networks against failure. Sci. Rep. **6**, 30521 (2016)
16. T. Morstyn, B. Hredzak, V.G. Agelidis, Distributed cooperative control of microgrid storage. IEEE Trans. Power Syst. **30**(5), 2780–2789 (2015)
17. P. Yi, Y. Hong, F. Liu, Initialization-free distributed algorithms for optimal resource allocation with feasibility constraints and application to economic dispatch of power systems. Automatica **74**, 259–269 (2016)
18. L. Ding, Q.-L. Han, B. Ning, D. Yue, Distributed resilient finite-time secondary control for heterogeneous battery energy storage systems under denial of service attacks. IEEE Trans. Ind. Inform. **16**(7), 4909–4919 (2020)
19. X. Lu, X. Yu, J. Lai, J.M. Guerrero, H. Zhou, Distributed secondary voltage and frequency control for islanded microgrids with uncertain communication links. IEEE Trans. Ind. Inform. **13**(2), 448–460 (2017)

Chapter 7
Distributed Optimal Control of DC Microgrids with Communication Delays

This chapter is concerned with load sharing and voltage regulation in DC microgrids. In order to address this issue, a multi-objective optimization problem with tunable weighting coefficients is first formulated for DC microgrids. Second, a distributed control scheme, which only requires local communications among neighbors, is proposed to solve the optimization problem. It is shown that, the proposed distributed control scheme has remarkable advantages in improving reliability and scalability of microgrids. Third, the distributed control scheme is extended to accommodate a constant communication delay. The effect of the communication delay on the stability of microgrids is explicitly characterized. Finally, the performance of the proposed control schemes is evaluated by a modified 6-bus microgrid with DC-powered trolleybus systems, in terms of their convergence, robustness to load variations, plug-and-play functionality, trade-off ability and effects of communication delays.

7.1 Problem Formulation

7.1.1 DC Micriogrid Models

Consider a DC microgrid consisting of n distributed energy resource (DER) units and n loads. The DC microgrid is modeled as a bidirectional and connected graph $\mathscr{G} = (\mathscr{V}, \mathscr{E})$, where $\mathscr{V} = \{1, 2, \ldots, n\}$ is the set of buses (nodes) and $\mathscr{E} \subseteq \mathscr{V} \times \mathscr{V}$ is the set of edges. An edge defined as $\varepsilon_{ij} \triangleq (i, j) \in \mathscr{E}$ implies that nodes i and j are connected by a DC transmission line with an impedance of $R_{ij} > 0$ and $L_{ij} > 0$, and node j is called a neighbor of node i. Let $N_i \triangleq \{j | \varepsilon_{ij} \in \mathscr{E}\}$ be the set of neighbors of node i. $\mathscr{A} = [w_{ij}] \in \mathbb{R}^{n \times n}$ represents a weighted adjacency matrix, where $w_{ii} = 0$ for all i, and $w_{ij} = \frac{1}{R_{ij}}$ if $\varepsilon_{ij} \in \mathscr{E}$, otherwise, $w_{ij} = 0$. The degree matrix of graph $\mathscr{G}$ is $\mathscr{D} = \text{diag}\{\varpi_1, \varpi_2, \ldots, \varpi_n\}$, where the diagonal elements are represented as

L. Ding et al., *Distributed Control and Optimization of Networked Microgrids*, Power Systems, https://doi.org/10.1007/978-3-030-95029-3_7

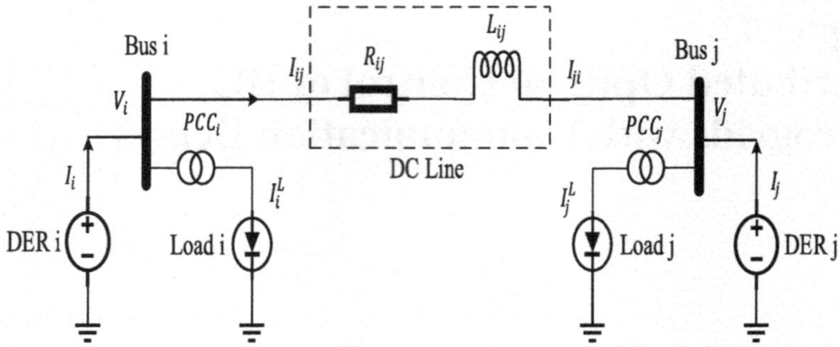

Fig. 7.1 Connection structure between DERs i and j in a DC microgrid

$\varpi_i = \sum_{j\in\mathcal{N}_i} w_{ij}$. The admittance matrix Y is defined as $Y = \mathcal{D} - \mathcal{A}$. A path is a sequence of connected edges in a graph. The graph $\mathcal{G}$ is said to be connected if there exists a path between any nodes. For a bidirectional and connected graph, the matrix Y is symmetric, i.e., $Y = Y^T$, and has a simple zero eigenvalue and all other eigenvalues are real and positive [1]. In addition, it is assumed that nodes can communicate with its physically connected neighbors. Hence, the communication topology is the same as the physical network.

The electrical structure between bus i and bus j connected by a DC transmission line is shown in Fig. 7.1. Each DER supplies power to a local DC load connected to the point of common coupling (PCC). It is assumed that each DER is a controllable current source with positive current injection $I_i > 0$ and each local load is a constant current load with current injection $I_i^L \geq 0$. Denote the voltage of bus i by V_i and the current of DC line (i, j) by I_{ij}. Without loss of generality, it is assumed that the directions of currents I_i, I_{ij} and I_i^L are given by Fig. 7.1. According to Kirchoff's current laws, one can know that the injection current I_i at node i is equal to the mount of outflow I_{ij} and I_i^L, namely,

$$I_i = \sum_{j\in N_i} I_{ij} + I_i^L.$$

It is assumed that the inductance L_{ij} is very small and even ignorable, which means that the transmission line is nearly resistive. Then, based on the QSL approximation method used in [2, 3], where it is assumed that $\frac{dI_{ij}}{dt} = 0$ since the inductance L_{ij} is enough small, one has

$$I_{ij} = \frac{V_i - V_j}{R_{ij}}, \quad I_{ji} = \frac{V_j - V_i}{R_{ji}}$$

Due to the fact that $R_{ij} = R_{ji}$, it is clear that $I_{ij} = -I_{ji}$. As a result, one has

$$I_i = \sum_{j \in N_i} \frac{V_i - V_j}{R_{ij}} + I_i^L. \tag{7.1}$$

Define $I \triangleq [I_1, I_2, \ldots, I_n]^T$, $V \triangleq [V_1, V_2, \ldots, V_n]^T$ and $I^L \triangleq [I_1^L, I_2^L, \ldots, I_n^L]^T$. By rewriting (7.1) into the compact form, the power flow model of microgrid can be modeled as

$$I = YV + I^L. \tag{7.2}$$

Since Y is a real-valued Laplacian matrix with $\mathbf{1}^T Y = \mathbf{0}$, one can get from (7.2) the current balance $\sum_{i \in \mathscr{V}} I_i = \sum_{i \in \mathscr{V}} I_i^L$.

7.1.2 *Objectives*

In the DC microgrid, DERs are used to cooperatively supply currents to loads on different buses. Since the current sources and the current loads are considered in this chapter, a basic concern is how to efficiently dispatch the current injection I_i at the DERs to all the loads I_i^L. One control policy for fair resource allocation is proportional load sharing, which is defined below.

Definition 7.1 Let I_i^s be the current at steady state for node i. Then the DC microgrid (7.2) is said to achieve proportional load sharing if the following conditions are satisfied

$$\frac{I_i^s}{m_i} = \frac{I_j^s}{m_j} = \mu \quad \text{for all } i, j \in \mathscr{V} \tag{7.3}$$

where $m_i > 0$ and $m_j > 0$ are the proportional coefficients and $\mu > 0$ is the ratio.

The parameters m_i for $i \in \mathscr{V}$ can be selected according to different practical requirements. For example, they are set as the capacities of DERs. According to the physical current balance constraint, it is clear that

$$\mu = \frac{\sum_{i=1}^n I_i^L}{\sum_{i=1}^n m_i} = \frac{\mathbf{1}^T I^L}{\sum_{i=1}^n m_i}.$$

In addition, another important performance criterion in a DC microgrid is to maintain desirable voltage profiles. Generally, to guarantee reliability of a DC microgrid, voltages at all nodes need to be regulated to be close to the rated values, i.e.

$$\lim_{t \to \infty} |V_i(t) - V_i^{ref}| \to \varepsilon, \quad i \in \mathscr{V} \tag{7.4}$$

where V_i^{ref} is the rated voltage value at node i and $\varepsilon > 0$ is a small constant.

In summary, the *objective* of this chapter is to design control schemes such that the DC microgrid achieves the proportional load sharing (7.3) and the voltage regulation (7.4) in a unified framework.

7.2 Global Optimization Schemes

In order to accommodate the control objective presented in Sect. 7.2, we introduce the following multi-objective optimization problem:

$$\min_{V_i, I_i} J = \frac{a}{2} \sum_{i=1}^{n} \left(\frac{I_i}{m_i} - \mu \right)^2 + \frac{b}{2} \sum_{i=1}^{n} (V_i - V_i^{ref})^2 \tag{7.5a}$$

$$\text{s.t.} \quad (7.2). \tag{7.5b}$$

where $a > 0$ and $b > 0$ are trade-off weighting coefficients satisfying $a + b = 1$.

Note that the performance index J in (7.5a) includes two terms: the first term is used to deal with the load (or current) sharing issue, and the second term is to drive the voltages at DERs to the reference values. As a result, the issues of the proportional load sharing (7.3) and the voltage regulation (7.4) can be handled by optimizing the performance index J. In general, it is difficult to achieve the proportional load sharing and the voltage regulation simultaneously due to the relationship (7.2). Hence, the weighting coefficients a and b are added in the performance index J to provide a tradeoff between the two conflicting objectives. Specially, if $a = 0$ and $b = 1$, the optimization problem (7.5) is reduced to the voltage regulation problem. Conversely, it is a pure proportional load sharing problem if $a = 1$ and $b = 0$. Both the cases are easily realized, thus they are not considered in this chapter.

Denote $V^{ref} \triangleq [V_1^{ref}, \ldots, V_n^{ref}]^T$ and $M \triangleq \text{diag}\{\frac{1}{m_i}\}$. Then, the optimal solution of (7.5) is derived as follows.

Theorem 7.1 *The global optimal solution of (7.5) is*

$$V^* = (aY^T MMY + b\mathscr{I}_n)^{-1}(bV^{ref} - aY^T MQI^L), \tag{7.6}$$

where $Q = M - \frac{1}{\mathbf{1}^T M^{-1} \mathbf{1}} \mathbf{1}\mathbf{1}^T$ *and* $\mathscr{I}_n$ *is the identity matrix.*

Proof It is clear that matrix Q is symmetric and $QY = MY$. In addition, it is easy to verify that $\mathbf{1}^T M^{-1} Q = 0$. Due to $\mathbf{1}^T I = \mathbf{1}^T I^L$ and μ, the performance index (7.5a) becomes

$$J = \frac{1}{2} a(MYV + QI^L)^T (MYV + QI^L) + \frac{1}{2} b(V - V^{ref})^T (V - V^{ref}) \tag{7.7}$$

Then, denoting the optimal voltage value by V^*, the optimality condition can be derived by

$$\left.\frac{\partial J}{\partial V}\right|_{V^*} = (aY^T MMY + b\mathscr{I}_n)V^* + aY^T MQI^L - bV^{ref}$$
$$= 0. \tag{7.8}$$

For a diagonal matrix $M > 0$, scalars $a > 0$ and $b > 0$, it is known that matrix $aY^T MMY + b\mathscr{I}_n$ is positive definite. Thus, it is easy to obtain (7.6) from (7.8). Correspondingly, the optimal current is calculated based on (7.2) and (7.6). The proof of Theorem 7.1 is complete. □

In Theorem 7.1, the optimal solution (V^*, I^*) of the optimization problem (7.5) is given. It is clearly seen from Theorem 7.1 that the optimal solution (7.6) is closely related to the weighting parameters a and b. By tuning the parameters a and b, one can obtain the different allocation of optimal solution (V^*, I^*), which explicitly reflects the trade-off between load sharing and voltage regulation. In fact, the selection of the parameters a and b is a specific and practical issue. In practice, the tuning parameters a and b are chosen based highly on specific needs. For example, if the load (or current) sharing is more critical, the parameter a needs to be increased, leading to that the parameter b is decreased due to the fact that $a + b = 1$. In this case, the optimal solution (7.6) plays more emphasis on the load sharing rather than the voltage regulation.

7.3 Distributed Optimization Schemes

In Sect. 7.2, the multi-objective optimization problem is formulated to deal with the load sharing and the voltage regulation and the optimal solution can be computed by using Theorem 7.1. Note that, however, in order to compute the optimal solution (7.6) a centralized controller is required to collect global information of the grid. In practical applications, this requirement is usually infeasible, especially, when the number of DERs is large and communication resources to support the central controller are limited. As a result, it is desirable to adopt a distributed control scheme that uses only local information but achieves the global optimal solution (7.6). Therefore, to achieve the optimization problem (7.5), a distributed control scheme is proposed as follows:

$$\dot{V}_i(t) = -k_1 a \sum_{j\in N_i} \frac{1}{R_{ij}} \left(\frac{\phi_i(t)}{m_i^2} - \frac{\phi_j(t)}{m_j^2} \right) - k_2 b \left(V_i(t) - V_i^{ref} \right), \tag{7.9a}$$

$$\dot{\phi}_i(t) = -\frac{k_1}{k_2}\phi_i(t) + \sum_{j\in N_i} \frac{1}{R_{ij}} \left(V_i(t) - V_j(t) \right) + I_i^L - \eta_i(t), \tag{7.9b}$$

$$\dot{\eta}_i(t) = -k_4 \sum_{j \in N_i} \frac{1}{R_{ij}} \left(\frac{\eta_i(t)}{m_i} - \frac{\eta_j(t)}{m_j} \right) - k_3(\eta_i(t) - I_i^L) + k_3\theta_i(t), \tag{7.9c}$$

$$\dot{\theta}_i(t) = -k_4 \sum_{j \in N_i} \frac{1}{R_{ij}} \left(\frac{\eta_i(t)}{m_i} - \frac{\eta_j(t)}{m_j} \right), \tag{7.9d}$$

where $\phi_i(t)$, $\eta_i(t)$ and $\theta_i(t)$ are auxiliary variables and $k_i > 0$ $(i = 1, 2, 3, 4)$ are control gains.

The distributed control scheme (7.9) is implemented through V_i at each node i. It is clear that the control updates of node i only depend on its own information with voltage and the auxiliary variables as well as those of its neighbors. In this scheme, the auxiliary variable ϕ_i for node i is used to estimate the error between the current I_i and its expected value $m_i\mu$. Note that the load sharing ratio μ involves the global information of all loads I_L^i and proportional coefficients m_i, thus it is unavailable to each node i. To address this issue, the proportional integral consensus-type control algorithm (7.9c) and (7.9d) is employed to have access to the global information μ for node i.

Lemma 7.1 *Let*

$$H_1 = \begin{bmatrix} k_2 b \mathscr{I}_n & k_1 a Y^T M M \\ -Y & \frac{k_1}{k_2} \mathscr{I}_n \end{bmatrix}$$

and $\lambda_p \triangleq \sigma_p(Y^T M M Y)$. *Then all the eigenvalues of* H_1 *are in the open right half plane. Moreover, the minimum real part of all the eigenvalues is* $\rho_1 \triangleq \min\{k_2 b, k_1/k_2\}$.

Proof It is straightforward to derive the conclusion of Lemma 7.1 by solving the characteristic equation of H_1. □

Lemma 7.2 *Given the initial conditions* $\theta_i(0) = 0$ $(i = 1, 2, \ldots, n)$, *the control schemes (7.9c) and (7.9d) converge to the equilibrium points* $(\bar{\eta}, \bar{\theta})$ *where* $\bar{\eta} \triangleq \mu M^{-1}\mathbf{1}$ *and* $\bar{\theta} \triangleq \mu M^{-1}\mathbf{1} - I^L$.

Proof Denoting

$$\eta \triangleq [\eta_1, \eta_2, \ldots, \eta_n]^T, \quad \theta \triangleq [\theta_1, \theta_2, \ldots, \theta_n]^T,$$
$$\tilde{\eta}(t) = \eta(t) - \bar{\eta}, \quad \tilde{\theta}(t) = \theta(t) - \bar{\theta},$$

one has the following error system

$$\dot{\tilde{\eta}}(t) = -k_3\tilde{\eta}(t) - k_4 Y M \tilde{\eta}(t) + k_3\tilde{\theta}(t), \tag{7.10a}$$
$$\dot{\tilde{\theta}}(t) = -k_4 Y M \tilde{\eta}(t). \tag{7.10b}$$

By setting $YM\tilde{\eta}(t) = 0$, the equilibrium point of (7.10b) is

$$M\bar{\tilde{\eta}} = c\mathbf{1},$$

where c is a constant. According to the definition of $\tilde{\theta}$, we have

$$\mathbf{1}^T\tilde{\theta}(0) = \mathbf{1}^T\bar{\theta} = \mathbf{1}^T M^{-1} Q = 0.$$

Since $\mathbf{1}^T Y = 0$, it is clear that $\mathbf{1}^T\dot{\tilde{\theta}}(t) = 0$, which implies that

$$\mathbf{1}^T\tilde{\theta}(t) = \mathbf{1}^T\tilde{\theta}(0) = 0.$$

From (7.10a), it follows that

$$\lim_{t\to\infty} \mathbf{1}^T\tilde{\eta}(t) = \lim_{t\to\infty} k_3\mathbf{1}^T\tilde{\theta}(t) = 0.$$

As a result, $\mathbf{1}^T\bar{\tilde{\eta}} = \mathbf{1}^T M^{-1}\mathbf{1}c = 0$, which means $c = 0$. The error system (7.10) can be rewritten by

$$\begin{bmatrix} \dot{\tilde{\eta}} \\ \dot{\tilde{\theta}} \end{bmatrix} = -H_2 \begin{bmatrix} \tilde{\eta} \\ \tilde{\theta} \end{bmatrix}$$

where

$$H_2 = \begin{bmatrix} k_3\mathscr{I}_n + k_4 YM & -k_3\mathscr{I}_n \\ k_4 YM & 0 \end{bmatrix}.$$

Similar to the method used in Lemma 7.1, the eigenvalues of matrix H_2 are calculated as 0, k_3 and $k_4\sigma_p(YM)$ $(p = 2, 3, \ldots, n)$. From [1], it is known that the zero eigenvalue ensures the states to reach consensus and is not related to the convergence rate. Since the Laplacian matrix Y is positive semi-definite and $M > 0$ is diagonal, it is easy to verify that $\sigma_p(YM) > 0$ $p = 2, 3, \ldots, n$ and $\sigma_2(YM)$ is its minimum. Therefore, it is concluded that $\tilde{\eta}(t) \to 0$ as $t \to \infty$. The proof of Lemma 7.2 is complete. □

We now state and establish the following main result.

Theorem 7.2 *For scalars $k_i > 0$ $(i = 1, 2, 3, 4)$, $a > 0$, $b > 0$ and the initial conditions $\theta_i(0) = 0$ $(i = 1, \ldots, n)$, the distributed control scheme (7.9) converges to the global optimal solution (7.6).*

Proof Together with $Y = Y^T$, the distributed control schemes (7.9a) and (7.9b) are rewritten in a compact form as

$$\dot{V}(t) = -k_1 a Y^T M M\phi(t) - k_2 b(V(t) - V^{ref}), \tag{7.11a}$$

$$\dot{\phi}(t) = -\frac{k_1}{k_2}\phi(t) + YV(t) + (I^L - \eta(t)), \tag{7.11b}$$

where $\phi(t) = [\phi_1(t), \phi_2(t), \ldots, \phi_n(t)]^T$. Then we calculate the equilibrium points $(\bar{\phi}, \bar{V})$ of (7.11) by

$$0 = -k_1 a Y^T M M \bar{\phi} - k_2 b(\bar{V} - V^{ref}), \tag{7.12a}$$

$$0 = -\frac{k_1}{k_2}\bar{\phi} + Y\bar{V} + I^L - \bar{\eta}. \tag{7.12b}$$

Based on Lemma 7.2, it is known that $\bar{\eta} = \mu M^{-1}\mathbf{1}$. From (7.12b), one gets

$$\bar{\phi} = \frac{k_2}{k_1}(Y\bar{V} + M^{-1}QI^L).$$

Then, substituting it into (7.12a) yields (7.6), which implies that the equilibrium point $\bar{V}$ of (7.9) is identical to the global optimal solution V^* in (7.6).

To analyze the convergence, we denote the error variables by

$$\tilde{V}(t) = V(t) - V^*, \quad \tilde{\phi}(t) = \phi(t) - \bar{\phi}.$$

Then, we have

$$\dot{\tilde{V}}(t) = -k_1 a Y^T M M \tilde{\phi}(t) - k_2 b \tilde{V}(t), \tag{7.13}$$

$$\dot{\tilde{\phi}}(t) = -\frac{k_1}{k_2}\tilde{\phi}(t) + Y\tilde{V}(t) - \tilde{\eta}(t), \tag{7.14}$$

where both equations in (7.12) are used.

Defining

$$z(t) = [\tilde{V}(t), \tilde{\phi}(t), \tilde{\eta}(t), \tilde{\theta}(t)]^T,$$

we have

$$\dot{z}(t) = -Hz(t),$$

where

$$H = \begin{bmatrix} H_1 & H_{12} \\ 0 & H_2 \end{bmatrix}, \quad H_{12} = \begin{bmatrix} 0 & 0 \\ 0 & \mathcal{I}_n \end{bmatrix}.$$

Since matrix H is block triangular, the stability of matrix H depends on matrices H_1 and H_2. Based on Lemmas 7.1 and 7.2, it is concluded that $z(t)$ converges to 0. Theorem 7.2 is proved. □

In Theorem 7.2, it is claimed that the global optimal solution (7.6) can be achieved by using the distributed control scheme (7.9). It is very beneficial to practical implementation since the centralized controller is no longer necessary.

7.4 Distributed Control Schemes in the Presence of Communication Delays

In Sect. 7.3, a distributed control strategy which uses its own and its neighbors' information is proposed to ensure that the system can converge to the global optimal solution. The study is based on ideal communication channels. However, information exchange is usually subject to communication delays due to unreliable communication channels. Therefore, in this section, we will analyze the impact of communication delays on the distributed control scheme.

It is assumed that the communication delay between neighbors is constant and denoted by $\tau > 0$. Then, the following distributed control scheme is presented

$$\dot{V}_i(t) = -k_1 a \sum_{j\in N_i} \frac{1}{R_{ij}} \left(\frac{\phi_i(t-\tau)}{m_i^2} - \frac{\phi_j(t-\tau)}{m_j^2} \right) - k_2 b\big(V_i(t) - V_i^{ref}\big), \tag{7.15a}$$

$$\dot{\phi}_i(t) = -\frac{k_1}{k_2}\phi_i(t) + \sum_{j\in N_i} \frac{1}{R_{ij}} \big(V_i(t-\tau) - V_j(t-\tau)\big) + (I_i^L - \eta_i(t)), \tag{7.15b}$$

$$\dot{\eta}_i(t) = -k_4 \sum_{j\in N_i} \frac{1}{R_{ij}} \left(\frac{\eta_i(t-\tau)}{m_i} - \frac{\eta_j(t-\tau)}{m_j} \right) - k_3(\eta_i(t) - I_i^L) + k_3\theta_i(t), \tag{7.15c}$$

$$\dot{\theta}_i(t) = -k_4 \sum_{j\in N_i} \frac{1}{R_{ij}} \left(\frac{\eta_i(t-\tau)}{m_i} - \frac{\eta_j(t-\tau)}{m_j} \right), \tag{7.15d}$$

Remark 7.1 It is worth mentioning that the control algorithm (7.15) uses the voltage measurement $V_i(t)$ and its delayed one $V_i(t-\tau)$ at node i. Due to the presence of the communication delay, $V_i(t-\tau)$ is coupled with its delayed neighbors' voltages $V_j(t-\tau)$. A reasonable physical explanation for this coupling is to compute the corresponding delayed link current between node i and j.

We now state and establish the following result.

Theorem 7.3 *For scalars $k_i > 0$ $(i = 1, 2, 3, 4)$, $a > 0$ and $b > 0$, let*

$$w_{1,n} = \sqrt{\frac{-\alpha + \sqrt{\alpha^2 - 4\beta_n}}{2}},$$

where

$$\alpha = k_2^2 b^2 + \frac{k_1^2}{k_2^2}, \quad \beta_n = k_1^2 b^2 - k_1^2 a^2 \lambda_n^2$$

and $\lambda_n > \frac{b}{a}$ is defined in Lemma 7.1. Then under the initial condition $\theta_i(0) = 0$, the distributed control scheme (7.15) asymptotically converges to the global optimal solution (7.6) for $\tau \in [0, \bar{\tau})$, where

$$\bar{\tau} \triangleq \begin{cases} \min\{\tau_{1,n}, \tau_{2,n}\}, & \lambda_n > \dfrac{b}{a} \\ \tau_{2,n}, & \lambda_n \leq \dfrac{b}{a} \end{cases}$$

with

$$\tau_{1,n} = \frac{1}{2w_{1,n}} \arctan\left\{\frac{\left(k_2 b + \frac{k_1}{k_2}\right) w_{1,n}}{w_{1,n}^2 - k_1 b}\right\}$$

$$\tau_{2,n} = \frac{\pi}{2k_4\sigma_n(YM)}.$$

Proof The compacted error system of (7.15) is written as

$$\dot{z}(t) = \hat{H}_1 z(t) + \hat{H}_2 z(t-\tau) \tag{7.16}$$

where

$$\hat{H}_1 = \begin{bmatrix} -k_2 b \mathscr{I}_n & 0 & 0 & 0 \\ 0 & -\frac{k_1}{k_2}\mathscr{I}_n & -\mathscr{I}_n & 0 \\ 0 & 0 & -k_3\mathscr{I}_n & k_3\mathscr{I}_n \\ 0 & 0 & 0 & 0 \end{bmatrix}$$

$$\hat{H}_2 = \begin{bmatrix} 0 & -k_1 a Y^T M M & 0 & 0 \\ Y & 0 & 0 & 0 \\ 0 & 0 & -k_4 Y M & 0 \\ 0 & 0 & -k_4 Y M & 0 \end{bmatrix}$$

In order to analyze the stability of (7.16), we first have its characteristic polynomial

$$\det(s\mathscr{I}_n - \hat{H}_1 - \hat{H}_2 e^{-\tau s}) = \prod_{p=1}^{n} f_p(s,\tau) \prod_{q=1}^{n} g_q(s,\tau),$$

where $f_p(s,\tau)$ and $g_q(s,\tau)$ are quasi-polynomials given by

$$f_p(s,\tau) = s^2 + \left(k_2 b + \frac{k_1}{k_2}\right) s + k_1 b + k_1 a \lambda_p e^{-2\tau s}$$

$$g_q(s,\tau) = s^2 + k_3 s + e^{-\tau s} k_4 \sigma_q(YM)(s + k_3)$$

From Theorem 7.2, it follows that (7.16) is stable if $\tau = 0$. Then, it is concluded that the error system (7.16) is asymptotically stable, i.e., $z(t) \to 0$ as $t \to \infty$ if $f_p(jw, \tau) \neq 0$ and $g_q(jw, \tau) \neq 0$ are satisfied for all $w, \tau \in [0, \bar{\tau})$, $p, q = 2, \ldots, n$. The upper bound $\bar{\tau}$ is a minimum value of the solutions τ to $f_p(jw, \tau) = 0$ or $g_q(jw, \tau) = 0$. In what follows, we will consider two cases:

Case I: Let $w_{1,p} > 0$ and $\tau_{1,p} > 0$ be the solutions of $f_p(jw, \tau) = 0$. Here, since the imaginary roots for $f_p(jw, \tau)$ are complex conjugate pairs, we only consider $w_{1,p} > 0$. By using Euler's formula

$$e^{-2\tau_{1,p}w_{1,p}j} = \cos(2\tau_{1,p}w_{1,p}) - j\sin(2\tau_{1,p}w_{1,p}),$$

we have

$$-w_{1,p}^2 + k_1 b + k_1 a \lambda_p \cos(2\tau_{1,p}w_{1,p}) = 0$$
$$\left(k_2 b + \frac{k_1}{k_2}\right) w_{1,p} - k_1 a \lambda_p \sin(2\tau_{1,p}w_{1,p}) = 0,$$

which are equivalent to

$$\sin(2\tau_{1,p}w_{1,p}) = \frac{(k_2 b + \frac{k_1}{k_2})w_{1,p}}{k_1 a \lambda_p}$$
$$\cos(2\tau_{1,p}w_{1,p}) = \frac{w_{1,p}^2 - k_1 b}{k_1 a \lambda_p}.$$

Hence, it is easy to obtain that

$$\tau_{1,p} = \frac{1}{2w_{1,p}}(k\pi + \arctan\left\{\frac{\left(k_2 b + \frac{k_1}{k_2}\right) w_{1,p}}{w_{1,p}^2 - k_1 b}\right\})$$

where k is the minimum integer satisfying $\tau_{1,p} > 0$. Due to the fact that $\sin^2(2\tau_{1,p}w_{1,p}) + \cos^2(2\tau_{1,p}w_{1,p}) = 1$, one has

$$(k_2 b + \frac{k_1}{k_2})^2 w_{1,p}^2 + (w_{1,p}^2 - k_1 b)^2 = k_1^2 a^2 \lambda_p^2.$$

Denoting $y_{1,p} = w_{1,p}^2 > 0$ and rearranging this equation result in

$$y_{1,p}^2 + \alpha y_{1,p} + \beta_p = 0. \tag{7.17}$$

Since $\alpha > 0$, we only need to examine the existence of positive solutions to (7.17) by β_p. If $\lambda_n > \frac{b}{a}$, there is at least one $\beta_p < 0$. Thus, one has the solutions to (7.17) as

$$y_{1,p} = \frac{-\alpha + \sqrt{\alpha^2 - 4\beta_p}}{2}, \tag{7.18}$$

where the solutions $y_{1,p} \leq 0$ are abandoned. For λ_n, we have the maximum $w_{1,n} = \sqrt{y_{1,n}}$. Then, it is easy to verify that $\tau_{1,n}$ is the minimum of $\tau_{1,p}$. If $\lambda_n \leq \frac{b}{a}$, then $\beta_p \geq 0$. It implies that there exist no positive solutions to (7.17).

Case II: Denote by $w_{2,q} > 0$ and $\tau_{2,q} > 0$ the solutions of $g_q(jw, \tau) = 0$. Similar to the lines of Case I, one can have

$$w_{2,q}^2 - k_4\sigma_q(YM)(k_3\cos(\tau_{2,q}w_{2,q}) - w\sin(\tau_{2,q}w_{2,q})) = 0,$$
$$k_3w_{2,q} + k_4\sigma_q(YM)(w\cos(\tau_{2,q}w_{2,q}) - k_3\sin(\tau_{2,q}w_{2,q})) = 0.$$

Then, it is easily calculated that

$$\sin(\tau_{2,q}w_{2,q}) = \frac{w_{2,q}}{k_4\sigma_q(YM)}, \qquad \cos(\tau_{2,q}w_{2,q}) = 0.$$

Thus, one has

$$\tau_{2,q} = \frac{\pi}{2k_4\sigma_q(YM)}. \tag{7.19}$$

It is easily verified that $\tau_{2,n}$ is the minimum for all $q = 2, \ldots, n$.

Based on Cases I and II, the delay margin $\bar{\tau}$ is given by choosing the minimum value of $\tau_{1,n}$ and $\tau_{2,n}$. The proof of Theorem 7.3 is complete. □

Remark 7.2 When a constant communication delay is involved in (7.15), the delay margin can be calculated by Theorem 7.3. In many practical applications, however, the communication delays between neighbors might be time-varying and nonuniform due to protocol variations in the nature of data transfer or uncertainties and disturbances in communication channels, thus leading to that the results derived in Theorem 7.3 will be no longer applicable. In this case, one of effective solutions is to employ a Lyapunov-Krasovskii functional method for dealing with the time-varying and nonuniform communication delays [4, 5]. Based on this method, a maximum allowable upper bound of the delay will be obtained by solving a set of linear matrix inequalities. For brevity, the detailed analysis and the related results are not provided in this chapter. Instead, a simulation case will be conducted in the Case 7.5.6 to show the effects of time-varying and nonuniform communication delays on the stability of microgrids.

7.5 Simulation Case Studies

7.5.1 *A Modified 6-Bus DC Microgrid for a Trolleybus System*

To illustrate the effectiveness of the proposed distributed control schemes, we consider a power supply network of the new Beijing Dual-Source Trolleybus System in [6] as a test system, whose configuration is shown in Fig. 7.2. Different from most existing trolleybus systems with uncontrolled rectifier units, the new system introduces controllable voltage source PWM rectifiers for supplying DC power to feeder lines. Such implementation provides several outstanding features:

(i) the power station is distributed rather than centralized, which can significantly reduce system power losses;
(ii) the power output of each converter is controllable and can be efficiently adjusted according to the information of neighboring PWM rectifiers.

As shown in Fig. 7.2, the vehicle line is divided into segments powered by the feeder lines. 4–8 power segments are connected to a power station whose supply radius is about 2 km. Each segment is not only physically connected to its neighbors but also can communicate with them via a communication network. In this system, both the physical and communication networks are identical and bidirectional. Based on the

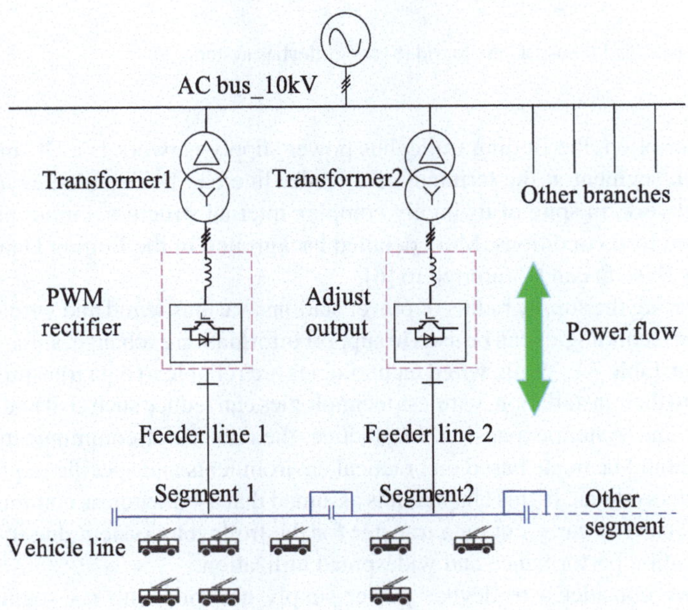

Fig. 7.2 Configuration of the trolleybus power supply network [6]

Table 7.1 Communication technologies for the trolley system [7]

Type	Technology	Data rate	Coverage	Frequency
Wired	PLC	2–45 Mbps	1–3 km	1–30 MHz
	Fiber-optical	10–1600 Gbps	Unlimited	3×10^8 GHz
Wireless	WiMAX	Up to 75 Mbps	1–5 km	2.5 GHz
	4G	Up to 1 Gbps	10 km	700–2500 MHz
	Micowave	155 Mbps	60 km	2–40 GHz

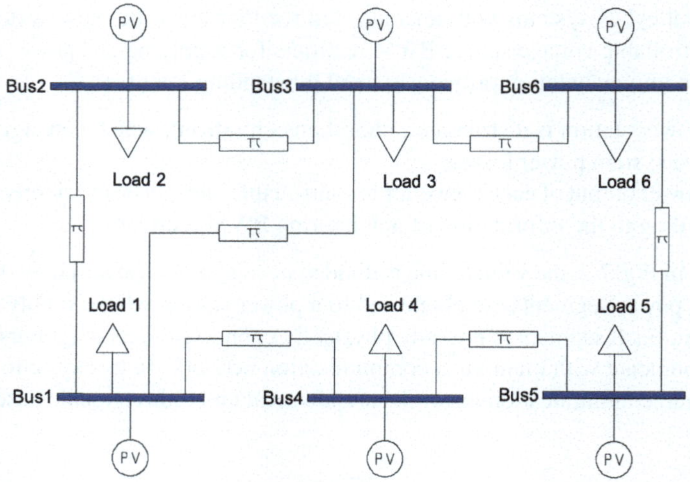

Fig. 7.3 A modified 6-bus DC microgrid [8] for trolleybus systems

above description, the Beijing trolleybus power supply network is a DC microgrid, where each segment at the terminal of a feeder line can be regarded as a node of the DC network, in spite of its highly complex internal structure, circuit protection systems and load conditions. More detailed background of the Beijing Dual-Source Trolleybus System can be referred to [6].

Considering the supply radius of power stations, various wired and wireless communication technologies can be used to support information exchange, some of which are listed in Table 7.1. While wired technologies are reliable to data transmission but are costly to their installation, wireless technologies can reduce such cost and improve scalability and system expansion. In practice, the choices of communication technologies should be made based on practical environments and specific requirements of trolleybus systems. In this chapter, it is assumed that a 4G wireless communication network is used to carry out data transfer for the trolleybus system due to its good communication performance and widespread utilization.

Here, we consider a trolleybus power supply network with six segments. As shown in Fig. 7.3, a modified 6-bus microgrid [8] is used to model the trolleybus power network, where generators, loads and buses represent feeder lines, vehicles and segments, respectively. All the simulation tests are performed by using Matlab/PSAT. The related parameters come from [6], namely, the resistance density of the transmission line is $0.2\Omega/\text{km}$ and the resistances among nodes are $R_{12} = R_{21} = 0.40\Omega$, $R_{13} = R_{31} = 0.37\Omega$, $R_{14} = R_{41} = 0.34\Omega$, $R_{23} = R_{23} = 0.38\Omega$, $R_{36} = R_{62} = 0.4\Omega$, $R_{45} = R_{45} = 0.31\Omega$ and $R_{56} = R_{56} = 0.36\Omega$, respectively. Let the proportional parameters $m_1 = m_2 = 1.500$, $m_3 = m_4 = 1.750$ and $m_5 = m_6 = 2.000$, which are related to the output capacities of the feeder lines. The nominal voltages of the nodes are set as $V_i^{ref} = 380\,\text{V}$ for all i. The loads are $I_1^L = 713\,\text{A}$, $I_2^L = 783\,\text{A}$, $I_3^L = 1009\,\text{A}$, $I_4^L = 842\,\text{A}$, $I_1^L = 921\,\text{A}$, $I_6^L = 803\,\text{A}$, respectively. Let the initial voltages $V_1(0) = 360\,\text{V}$, $V_2(0) = 380\,\text{V}$, $V_3(0) = 400\,\text{V}$, $V_4(0) = 375\,\text{V}$, $V_5(0) = 366\,\text{V}$, $V_6(0) = 350\,\text{V}$. Set the control gains $k_1 = 3, k_2 = 7$, $k_3 = 10$ and $k_4 = 10$ for all the following cases.

7.5.2 Optimality

Let the weighting coefficients $a = 0.5$ and $b = 1 - a = 0.5$. By Theorem 7.1, the global optimal solution is computed as

$$\begin{aligned}
&V_1^* = 371.51V,\ V_2^* = 358.48\,\text{V},\ V_3^* = 359.81\,\text{V},\\
&V_4^* = 385.23\,\text{V},\ V_5^* = 398.29\,\text{V},\ V_6^* = 406.67\,\text{V};\\
&I_1^* = 736.88\,\text{A},\ I_2^* = 746.91\,\text{A},\ I_3^* = 863.70\,\text{A},\\
&I_4^* = 840.22\,\text{A},\ I_5^* = 939.85\,\text{A},\ I_6^* = 943.44\,\text{A}.
\end{aligned} \tag{7.20}$$

Next, we demonstrate the effectiveness of the distributed control scheme (7.9), i.e., it can converge to the global optimal solution (7.20). Here, it is assumed that the communication network is ideal without communication delays. Based on the distributed control scheme (7.9), the evolutions of the voltages and currents are plotted in Fig. 7.4.

From Fig. 7.4, it is clear that the output voltages and currents under the distributed control scheme (7.9) are identical to the global solution (7.20).

7.5.3 The Impacts of Load Variations

In a trolleybus supply network, the load demand at each segment may experience frequent changes due to penetration or removal of vehicles. Thus, it is necessary to evaluate robustness of the proposed distributed control algorithm to load changes. Suppose that the load demand at bus 3 is reduced by 30% at $t = 10\,\text{s}$, and then is restored after $t = 20\,\text{s}$. Using the control algorithm (7.9) with $a = b = 0.5$, we plot the evolutions of voltages and currents at all nodes in Fig. 7.5. It is observed from Fig. 7.5 that, when the load variations occur at $t = 10s$ and $20s$, the control

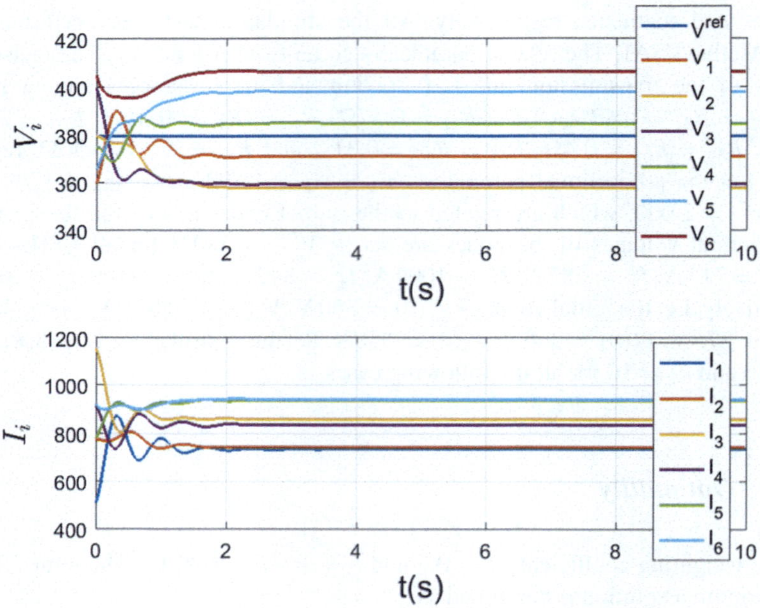

Fig. 7.4 Evolutions of voltages and currents under distributed control scheme (7.9) with $a = 0.5$ and $b = 0.5$

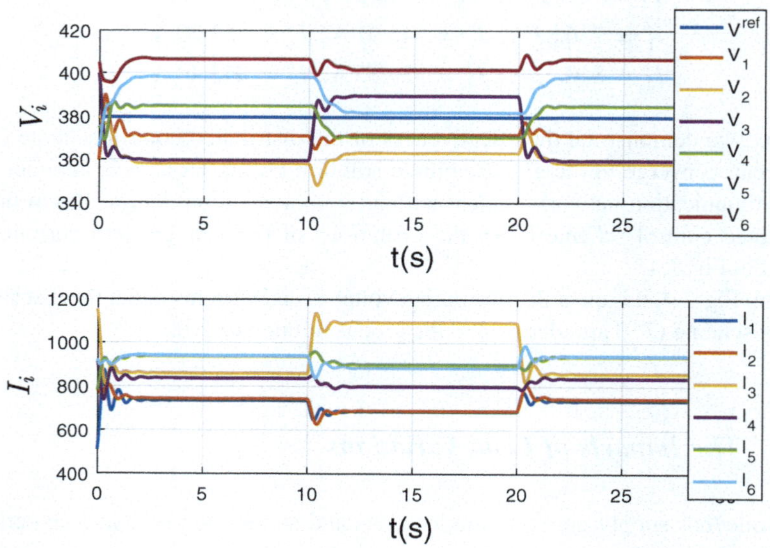

Fig. 7.5 Impacts of load variation on voltages and currents

algorithm (7.9) can take a prompt response such that the voltage and current outputs are correspondingly adjusted. Thus, it is verified that the distributed control algorithm (7.9) is robust to load variations.

7.5.4 Plug-and-Play Features

Due to intermittent nature of DERs, the feature of plug and play becomes very critical in a microgrid. For example, one segment in the trolleybus system may be plugged in or out due to maintenance services or system reconstruction. Next, we will demonstrate plug and play ability of the distributed control algorithm (7.9) with $a = b = 0.5$. Initially, one segment (named node 7) is operated independent of other nodes. So the voltage and current at node 7 equal to its initial voltage and the load current, respectively. At $t = 10$ s, it is connected to both nodes 5 and 6 with $R_{57} = 0.4\Omega$, $R_{67} = 0.35\Omega$, $I_7^L = 700$ A and $m_i = 1.000$, respectively. Then, it is again unplugged from the trolleybus systems at $t = 20$ s. The evolutions of voltage and current outputs are shown in Fig. 7.6. It is observed from Fig. 7.6 that the outputs of voltages and currents are changed for $t \in [10s, 20s)$ since node 7 is connected to the trolleybus system, which implies that the distributed control algorithm (7.9) shows good adaptability for plugging in and out.

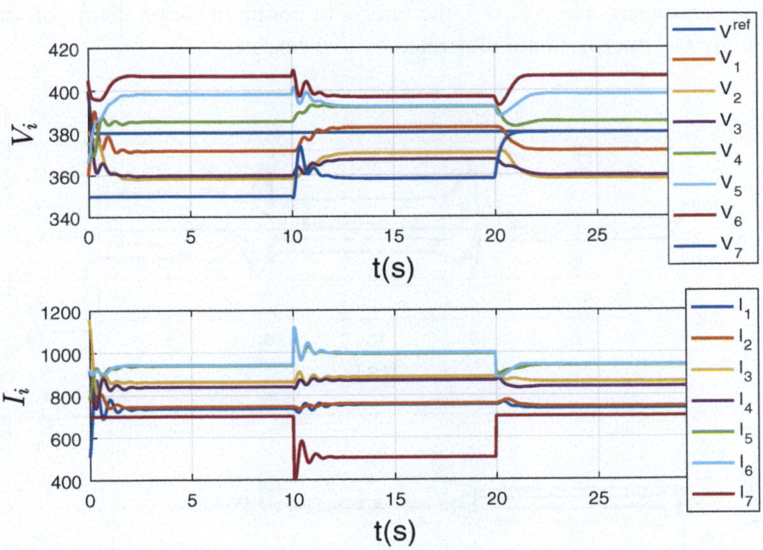

Fig. 7.6 Evolutions of voltages and currents under plug-and-play operation

7.5.5 *The Tradeoff Between Load Sharing and Voltage Regulation*

As stated previously, one cannot achieve perfect load sharing and voltage regulation simultaneously. But one can achieve a desirable trade-off by tuning weighting coefficients a and b. For example, when $a \to 0$, it means that the control scheme will focus only on voltage management, while load sharing will be almost ignored. When $a \to 1$, it mainly deals with load sharing rather than voltage regulation. In order to show the effects of weighting coefficient tuning on the optimal solution, we give a comprehensive simulation case: (i) $a = 0.10$ for $t \in [0, 10s)$; (ii) $a = 0.50$ for $t \in [10\,\text{s}, 20\,\text{s})$; (iii) $a = 0.09$ for $t \in [20\,\text{s}, 30\,\text{s}]$. The evolutions of all nodes' voltages and currents are depicted in Fig. 7.7. It is shown that, for $t \in [0, 10\,\text{s})$ each voltage V_i is mostly regulated to the nominal value 380 V, but it causes underperformance in the current sharing (I_i/m_i). For $t \in [20\,\text{s}, 30\,\text{s}]$, the currents I_i/m_i nearly reach the desirable consensus while the voltage regulation is not well performed. This implies that there is indeed a tradeoff between the load sharing and the voltage regulation.

7.5.6 *The Effects of Communication Delays*

With the parameters $a = b = 0.5$, the effects of communication delays on the stability of the DC microgrid are illustrated by two cases:

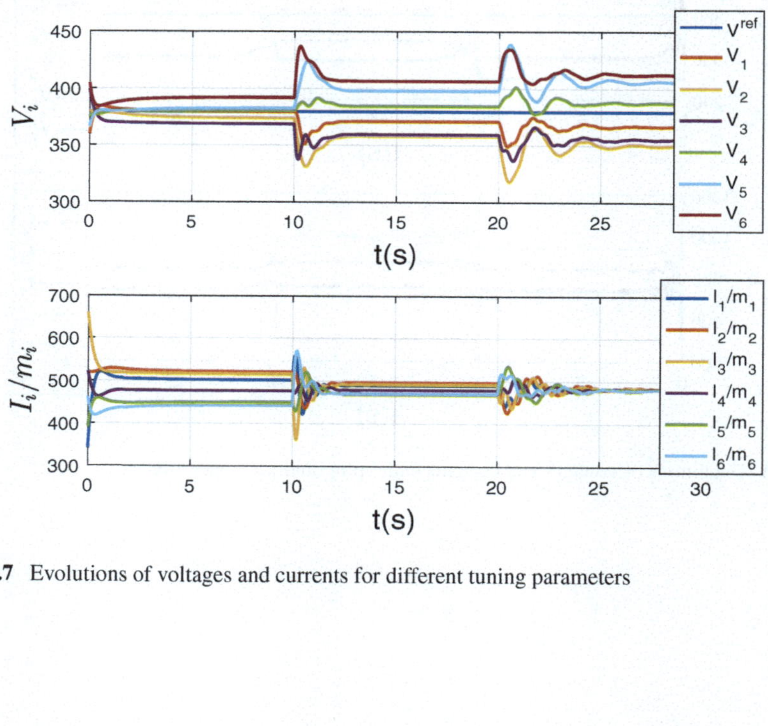

Fig. 7.7 Evolutions of voltages and currents for different tuning parameters

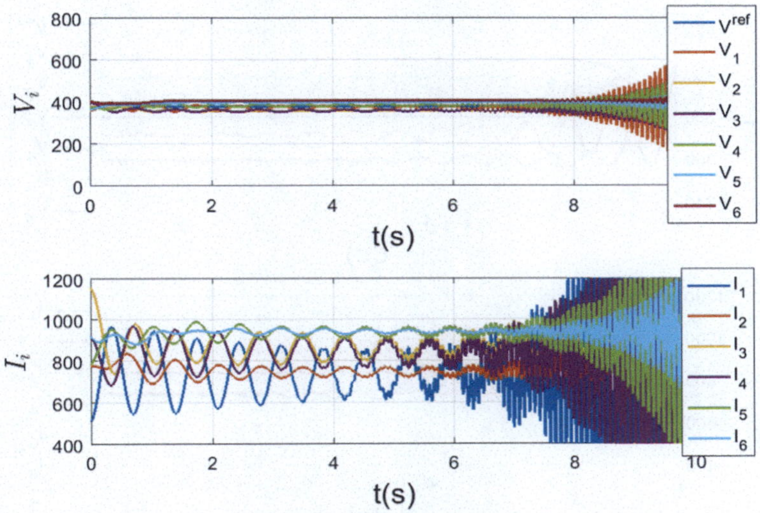

Fig. 7.8 Evolutions of voltages and currents under control scheme (7.15) with $\tau = 0.021s > \bar{\tau}$

(1) Constant communication delay: Based on Theorem 7.3, we compute $\tau_{1,n} = 0.0237$ s and $\tau_{2,n} = 0.0209$ s. Thus, the upper bound of communication delay is $\bar{\tau} = \min\{\tau_{1,n}, \tau_{2,n}\} = 0.0209$ s. Now, let $\tau = 0.021s > \bar{\tau}$, and we plot the outputs of voltages and currents in Fig. 7.8. From this figure, it is apparent that the distributed control scheme (7.15) cannot achieve the global optimal solution (7.20), and is in fact unstable. On the other hand, by setting $\tau = 0.017\,\text{s} < \bar{\tau}$, the evolutions of the voltages and currents are shown in Fig. 7.9, demonstrating that the distributed control scheme (7.15) converges to the global optimal solution (7.20). Therefore, these results validate the conclusion of Theorem 7.3.

In order to characterize the effects of communication delay on control performance, we define the control accuracy as $\varrho = |J - J^*|$. Then, Table 7.2 gives the convergence times under different communication delays and control accuracy. According to Table 7.2, it is clear that the convergence times of the distributed control scheme (7.15) become longer as the communication delay increases. In conclusion, the presence of communication delays degrades the convergence rate and the control performance.

(2) Time-varying communication delay: In this case, we consider the time-varying and nonuniform communication delays, which are denoted by $\tau_{ij}(t)$. It is assumed that $\tau_{ij}(t) \in [0.005\,\text{s}, 0.015\,\text{s}]$ is randomly generated. Then, we depict the evolution of voltages and currents in Fig. 7.10, from which it is clearly seen that the distributed control algorithm (7.15) can still achieve the optimal solution (7.20) even if the time-varying and nonuniform communication delays are taken into account.

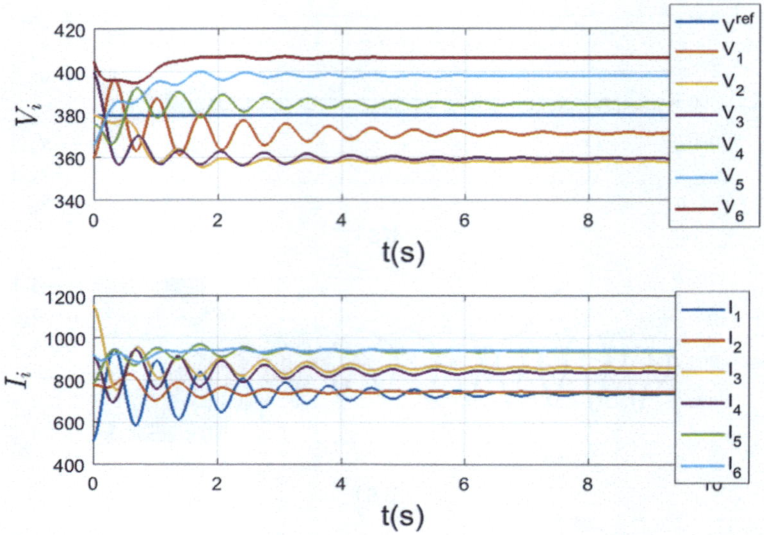

Fig. 7.9 Evolution of voltages and currents under distributed control scheme (7.15) with $\tau = 0.017s < \bar{\tau}$

Table 7.2 Comparison of Convergence times under different delays and accuracy requirements

$\varrho \tau$	0	0.005	0.010	0.015	0.017	0.020
10^{-1}	2.855	3.808	5.208	8.987	12.122	23.761
10^{-2}	3.543	4.511	6.517	10.726	14.550	28.641
10^{-3}	4.149	5.212	7.566	12.460	16.976	33.522

7.6 Concluding Remarks

In this chapter, a multi-objective optimization method has been developed to deal with the issues of proportional load sharing and voltage regulation for DC microgrids. To solve this optimization problem, a distributed control algorithm has been proposed. Theoretical results have shown that the equilibrium point of the distributed control algorithm is completely identical to the global optimal solution. Moreover, the distributed control algorithm has been extended to accommodate a constant communication delay. By an eigenvalue-based analysis approach, the maximum allowable upper bound of the communication delay has been computed. By the simulation results on a 6-bus microgrid of trolleybus systems, it has been verified that the proposed control algorithms possess good abilities in improving reliability and maintaining scalability, especially, in terms of their fast convergence, robustness to load variations and communication delays, plug-and-play feature and trade-off functionality.

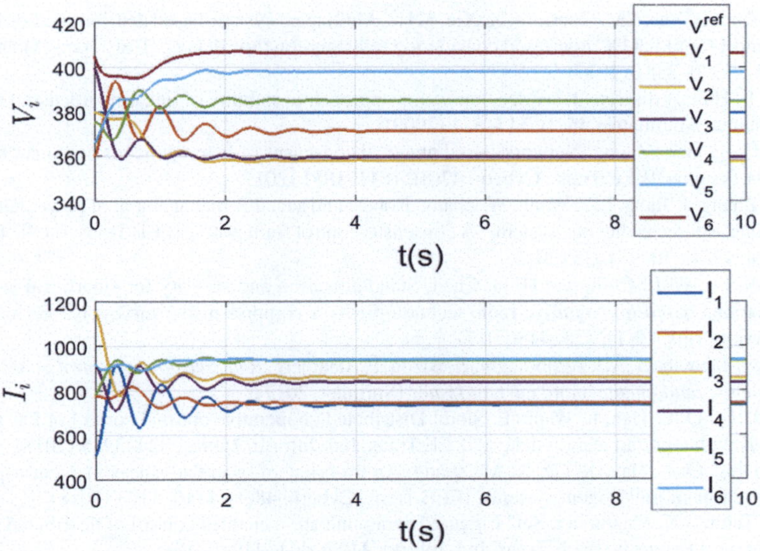

Fig. 7.10 Evolution of voltages and currents under distributed control scheme (7.15) with $\tau_{ij}(t) \in [0.005\text{ s}, 0.015\text{ s}]$

7.7 Notes

It should be pointed out that, the results in this chapter are based mainly on [9], which extends the research work in [6] to deal with distributed optimization of DC microgrids considering communication delays. Note that the objective function presented in this chapter does not include some important indices such as transmission line losses and operational limits of voltages and currents. Thus, our future research work will focus on distributed optimal power management and control of DC microgrids by taking into account these important indices. In addition, it would be a promising topic to develop distributed optimization algorithm incorporating event and self-triggered communication strategies [10, 11] such that the optimization solution can be obtained while reducing much consumption of communication resources.

References

1. R. Olfati-Saber, R. Murray, Consensus problems in networks of agents with switching topology and time-delays. IEEE Trans. Autom. Control **49**(9), 1520–1533 (2004)
2. M. Tucci, S. Riverso, J.C. Vasquez, J.M. Guerrero, G. Ferrari-Trecate, A decentralized scalable approach to voltage control of DC islanded microgrids. IEEE Trans. Control Syst. Technol. **24**(6), 1965–1979 (2016)

3. C. Dou, D. Yue, J.M. Guerrero, X. Xie, S. Hu, Multiagent system-based distributed coordinated control for radial DC microgrid considering transmission time delays. IEEE Trans. Smart Grid **8**(5), 2370–2381 (2017)
4. Q.-L. Han, A discrete delay decomposition approach to stability of linear retarded and neutral systems. Automatica **45**(2), 517–524 (2009)
5. L. Ding, W.X. Zheng, Network-based practical consensus of heterogeneous nonlinear multiagent systems. IEEE Trans. Cybern. **47**(8), 1841–1851 (2017)
6. D. Zhang, J. Jiang, L.Y. Wang, W. Zhang, Robust and scalable management of power networks in dual-source trolleybus systems: A consensus control framework. IEEE Trans. Intell. Transp. Syst. **17**(4), 1029–1038 (2016)
7. T. N. Le, W.-L. Chin, and H.-H. Chen, Standardization and security for smart grid communications based on cognitive radio technologies—A comprehensive survey, IEEE Commun. Surveys Tuts. **19**(1), 423–445(2017)
8. N.M. Tabatabaei, A.J. Aghbolaghi, N. Bizon, F. Blaabjerg, *Reactive power control in AC power systems: fundamentals and current Issues* (Springer, 2017)
9. L. Ding, Q.-L. Han, L. Wang, E. Sindi, Distributed cooperative optimal control of DC microgrids with communication delays. IEEE Trans. Ind. Inform. **14**(9), 3924–3935 (2018)
10. L. Ding, Q.-L. Han, X. Ge, X.-M. Zhang, An overview of recent advances in event-triggered consensus of multi-agent systems. IEEE Trans. Cybern. **48**(4), 1110–1123 (2018)
11. M. Tahir, S.K. Mazumder, Self-triggered communication enabled control of distributed generation in microgrids. IEEE Trans. Ind. Inform. **11**(2), 441–449 (2015)

Chapter 8
Distributed Energy Management for Smart Grids

This chapter is concerned with distributed energy management and control issues of both generators and loads, aiming to maximize the total social welfare which balances generation-side expanses, user-side payments, and transmission line costs. A distributed control strategy with continuous information exchange among neighbors is first proposed. It is shown that this distributed algorithm asymptotically achieves the global optimal power outputs on generators and the optimal electricity usage on loads. To reduce communication resource consumptions, the distributed optimization algorithm is further expanded to incorporate event-triggered communications and control mechanisms. In this algorithm, an event-triggered condition for each generator and each load are employed to determine when its related state information should be sampled and transmitted to its neighbors. Compared with the standard periodic sampling and communication schemes, this distributed event-triggered algorithm can significantly reduce communication data flows while achieving the nearly identical control performance to that under continuous data communications. The theoretical results are validated by using a simulation case study with distributed generators and multiple loads on an IEEE 9-bus system.

8.1 Preliminaries and Problem Formulation

8.1.1 Notations and Preliminaries

Some important properties on convex sets and projection are given, referring to [1]. Let Ω be a closed set. The set Ω is convex if for any $x_1, x_2 \in \Omega, \alpha x_1 + (1-\alpha)x_2 \in \Omega$, for all $0 < \alpha < 1$. Denote $\mathscr{C}_\Omega(x)$ as the normal cone of Ω at x. $\mathscr{T}_\Omega(x)$ represents the tangent cone of Ω at x. If the set Ω is convex, $\mathscr{T}_\Omega(x)$ is the polar cone to

L. Ding et al., *Distributed Control and Optimization of Networked Microgrids*, Power Systems, https://doi.org/10.1007/978-3-030-95029-3_8

$\mathscr{C}_\Omega(x)$, namely, $\mathscr{T}_\Omega(x) = \{y | \langle y, d\rangle \le 0, \forall d \in \mathscr{C}_\Omega(x)\}$. Denote the projection of x on a closed convex set Ω by $\mathscr{P}_\Omega(x) = \arg\min_{y\in\Omega} \|x - y\|$. For $\forall x \in \mathbb{R}^n$, $\forall y \in \Omega$, one has $\langle x - \mathscr{P}_\Omega(x), \mathscr{P}_\Omega(y) - y\rangle \ge 0$.

Lemma 8.1 (LaSalle's Invariant Set Theorem [2]) *Let $\Omega \subset D$ be a compact set that is positively invariant with respect to $\dot{x} = f(x)$, where $f : D \to \mathbb{R}^n$ is a locally Lipschitz map from a domain D to R^n. Let $V : D \to \mathbb{R}$ be a continuously differentiable function such that $\dot{V}(x) \le 0$ in Ω. Let E be the set of all points in Ω where $\dot{V}(x) = 0$. Let $\mathscr{M}$ be the largest invariant set in E. Then every solution starting in Ω approaches $\mathscr{M}$ as $t \to \infty$.*

8.1.2 Network Models

Consider an AC power grid consisting of n generators and m loads. There are two sets of buses: generator buses $\mathscr{N}_G = \{1, 2, \ldots, n\}$ and load buses $\mathscr{N}_L = \{n+1, n+2, \ldots, n+m\}$ with $\mathscr{N} = \mathscr{N}_G \cup \mathscr{N}_L$. The grid can be described by an undirected graph $\mathscr{G} = (\mathscr{V}, \mathscr{E})$, where $\mathscr{V} = \{1, 2, \ldots, n+m\}$ is the set of buses (nodes) and $\mathscr{E} \subseteq \mathscr{V} \times \mathscr{V}$ is the set of edges. An edge defined as $\varepsilon_{ij} \triangleq (i, j) \in \mathscr{E}$ implies that nodes i and j are connected by an AC transmission line with an admittance of magnitude $Y_{ij} > 0$ with $Y_{ij} = Y_{ji}$. Let $N_i \triangleq \{j | (i, j) \in \mathscr{E}\}$ be the set of neighbors of node i. $\mathscr{A} = [w_{ij}] \in \mathbb{R}^{n\times n}$ represents a weighted adjacency matrix, where $w_{ii} = 0$ for all i, and $w_{ij} = Y_{ij}$ if $\varepsilon_{ij} \in \mathscr{E}$, otherwise, $w_{ij} = 0$. The degree matrix of graph $\mathscr{G}$ is $\mathscr{D} = \text{diag}\{\varpi_1, \varpi_2, \ldots, \varpi_{n+m}\}$, where the diagonal elements are represented as $\varpi_i = \sum_{j\in\mathscr{C}_i} Y_{ij}$. Correspondingly, the Laplacian matrix of the graph $\mathscr{G}$ is $L = \mathscr{D} - \mathscr{A}$. A path is a sequence of connected edges in a graph. The graph $\mathscr{G}$ is said to be connected if there exists a path between any nodes. In addition, it is assumed that nodes can communicate with its physically connected neighbors. Hence, the communication topology is the same as the physical network.

Assumption 8.1.1 Graph $\mathscr{G}$ indicating both the physical network and communication topologies is undirected and connected.

The real power flow over line from bus i to bus j is computed by [3]

$$P_{ij} = V_i V_j Y_{ij} \sin(\theta_i - \theta_j), \quad i \in \mathscr{N} \tag{8.1}$$

where V_i and θ_i are the magnitude and phase angle of the voltage at node i. Then, the real power injection at bus i is given by

$$P_i = \sum_{j\in N_i} P_{ij} = \sum_{j\in N_i} V_i V_j Y_{ij} \sin(\theta_i - \theta_j), \quad i \in \mathscr{N} \tag{8.2}$$

To simplify the analysis, the DC approximation is widely used for computation of power flow [3], where it is assumed that the power lines are lossless, the phase angle difference $|\theta_i - \theta_j|$ is small and the voltage magnitude V_i at bus i is set as $1\,p.u$ for all $i \in \mathcal{N}$. Then, by a linear approximation that $\sin(\theta_i - \theta_j) \approx \theta_i - \theta_j$, the power flow (8.2) can be rewritten as

$$P_i = \sum_{j \in N_i} Y_{ij}(\theta_i - \theta_j), \ \ i \in \mathcal{N} \tag{8.3}$$

Without loss of generality, it is assumed that $P_i > 0$ for $i \in \mathcal{N}_G$ and $P_i < 0$ for $i \in \mathcal{N}_L$.

8.1.3 Social Welfare Maximization

In this chapter, we consider a competitive power market with generators, wholesalers, and customers. In such an energy market, there is a tendency that both the generation and demand are managed together. In other words, all of them become the participants in running the power system. To address the issue of improving overall costs and maintaining the fairness of all generators and loads, a total social welfare maximization problem is introduced in [4–6] as

$$\max J = \left[\sum_{i \in \mathcal{N}_G} W_{i,G}(P_{i,G}, p) + \sum_{r \in \mathcal{N}_L} W_{r,L}(P_{r,L}, p) \right], \tag{8.4a}$$

$$\text{s.t.} \quad \sum_{i \in \mathcal{N}_G} P_{i,G} = \sum_{r \in \mathcal{N}_L} P_{r,L}, \tag{8.4b}$$

$$P_{i,G}^m \le P_{i,G} \le P_{i,G}^M, \tag{8.4c}$$

$$P_{r,L}^m \le P_{r,L} \le P_{r,L}^M, \tag{8.4d}$$

where $P_{i,G}$ and $P_{r,L}$ are the generation power for $i \in \mathcal{N}_G$ and the consumption power for $r \in \mathcal{N}_L$, respectively; $W_{i,G}(\cdot)$ and $W_{r,L}(\cdot)$ denote the welfare functions of generation and demand sides, respectively; p is the price of electricity, $P_{i,G}^m$ and $P_{i,G}^M$ are the minimum and maximum generation power of generator $i \in \mathcal{N}_G$, respectively; $P_{r,L}^m$ and $P_{r,L}^M$ is the minimum and maximum demand power of load $r \in \mathcal{N}_L$, respectively.

Note that the maximum social welfare problem is subject to the power balance constraint (8.4b) and the operational constraints (8.4c–8.4d), when (small) power loss in the network is ignored. In this formulation, the wholesaler is viewed as a consumer advocator and grouped with the demand side, and hence is not given explicitly in the social welfare (8.4a).

(1) Generation welfare: Define the welfare function for the ith generator as

$$W_{i,G}(P_{i,G}, p) = pP_{i,G} - C_i(P_{i,G}) - \frac{\rho}{4} \sum_{j \in N_i} Y_{ij}(\theta_i - \theta_j)^2, \tag{8.5}$$

where $pP_{i,G}$ means the benefit of generator i with the selling price p, $C_i(\cdot)$ is the cost function of generator i and ρ is a tradeoff parameter which can be set according to a practical requirement.

The term $\frac{\rho}{4}\sum_{j\in N_i} Y_{ij}(\theta_i-\theta_j)^2 = \frac{\rho}{4}\sum_{j\in N_i} P_{ij}^2/Y_{ij}$ represents consolidated transmission line cost charged to the selling generator. This reflects costs on line maintenance, protection systems, communications, transformers, and wearing and aging, etc. The weight $1/Y_{ij}$ scales the cost to be proportional to the length of the line. The $C_i(\cdot)$ is expressed as a convex quadratic function

$$C_i(P_{i,G}) = a_i P_{i,G}^2 + b_i P_{i,G} + c_i, \tag{8.6}$$

where $a_i > 0$, b_i and c_i are cost parameters.

(2) Demand welfare: Define the demand welfare function for the rth consumer as

$$W_{r,L}(P_{r,L}, p) = U_r(P_{r,L}) - pP_{r,L} - \frac{\rho}{4}\sum_{j\in N_r} Y_{rj}(\theta_r-\theta_j)^2, \tag{8.7}$$

namely, the utility $U_r(P_{r,L})$ of consumer r minuses the payment $pP_{r,L}$ of consumer r to electricity providers and the transmission cost $\frac{\rho}{4}\sum_{j\in N_i} Y_{rj}(\theta_r-\theta_j)^2$ charged to the buyer. For each $r\in\mathcal{N}_L$, its consuming function is denoted by $U_r(P_{r,L})$, representing the satisfaction level of the electricity consumer, and it has the following properties[7]:

(i) $U_r(P_{r,L})$ is continuously differentiable and nondecreasing, i.e., $\frac{\partial U_r(P_{r,L})}{\partial P_{r,L}} \geq 0$;
(ii) The satisfaction level saturates at a point and $\frac{\partial^2 U_r(P_{r,L})}{\partial P_{r,L}^2} \leq 0$ before saturation;
(iii) $U_r(0) = 0$, indicating zero satisfaction without electricity.

In this chapter, to be concrete, we use the following utility function $U_r(P_{r,L})$ [5, 6]

$$U_r(P_{r,L}) = \beta_r P_{r,L} - \alpha_r P_{r,L}^2, \quad P_{r,L}^m \leq P_{r,L} \leq P_{r,L}^M \tag{8.8}$$

where β_r and $\alpha_r > 0$ are the parameters for the rth consumer.

According to $W_{i,G}(P_{i,G}, p)$ in (8.5) and $W_{r,L}(P_{r,L}, p)$ in (8.7), one has

$$\begin{aligned} J = &\sum_{r\in\mathcal{N}_L} U_r(P_{r,L}) - \sum_{i\in\mathcal{N}_G} C_i(P_{i,G}) - \frac{\rho}{4}\sum_{i\in\mathcal{N}}\sum_{j\in N_i} Y_{ij}(\theta_i-\theta_j)^2 \\ &+ p\left(\sum_{i\in\mathcal{N}_G} P_{i,G} - \sum_{r\in\mathcal{N}_L} P_{r,L}\right). \end{aligned}$$

The power flow model (8.3) implies that $P_{i,G} = P_i$ for $i\in\mathcal{N}_G$ and $P_{r,L} = -P_r$ for $r\in\mathcal{N}_L$. As a result, the power balance constraint (8.4b) is satisfied and redundant. Let

$$P \triangleq [P_{1,G}, \ldots, P_{n,G}, -P_{n+1,L}, \ldots, -P_{n+m,L}]^T$$
$$\theta \triangleq [\theta_1, \ldots, \theta_{n+m}]^T.$$

Denote the sets of generation power and demand power satisfying the box constraints by

$$\mathscr{F}_i^G \triangleq \{P_{i,G}, i \in \mathscr{N}_G | P_{i,G}^m \leq P_{i,G} \leq P_{i,G}^M\}$$
$$\mathscr{F}_r^L \triangleq \{P_{r,L}, r \in \mathscr{N}_L | P_{r,L}^m \leq P_{r,L} \leq P_{r,L}^M\},$$

respectively. Let $\mathscr{F} \triangleq \mathscr{F}_1^G \times \cdots \times \mathscr{F}_n^G \times \mathscr{F}_{n+1}^L \times \cdots \times \mathscr{F}_{n+m}^L$ be the Cartesian product of $\mathscr{F}_i^G$ and $\mathscr{F}_r^L$. It is clear that the sets $\mathscr{F}_i^G$, $\mathscr{F}_r^L$ and $\mathscr{F}$ are convex [1]. Then, the SWM problem (8.4) is equivalent to the following constrained minimization problem.

$$\min_{P \in \mathscr{F}, \theta} J_1 = \sum_{i \in \mathscr{N}_G} C_i(P_{i,G}) - \sum_{r \in \mathscr{N}_L} U_r(P_{r,L}) + \frac{\rho}{4} \sum_{i \in \mathscr{N}} \sum_{j \in N_i} Y_{ij}(\theta_i - \theta_j)^2, \tag{8.9a}$$

$$\text{s.t.} \quad P_{i,G} = \sum_{j \in N_i} Y_{ij}(\theta_i - \theta_j), \ i \in \mathscr{N}_G, \tag{8.9b}$$

$$P_{r,L} = -\sum_{j \in N_r} Y_{rj}(\theta_r - \theta_j), \ r \in \mathscr{N}_L, \tag{8.9c}$$

$$P_{i,G} \in \mathscr{F}_i^G, \ P_{r,L} \in \mathscr{F}_r^L. \tag{8.9d}$$

Remark 8.1 In the network model (8.3), the generator and load buses are interpreted according to their net power injections into the power network. All generators and loads connected to a bus are first grouped and their total net power injection is calculated. Bus i is called a "generator bus" if its net injection $P_i > 0$, and a "load bus" if the net injection $P_i < 0$. If the net power $P_i(t) = 0$, then bus i is not a participant in the problem. The net powers of the buses are managed in our optimization problem (8.9).

8.2 Distributed Optimal Solutions

In this section, a distributed control algorithm with continuous communication will be presented. Then the convergence analysis will be conducted.

8.2.1 Distributed Algorithms

Suppose that generators and loads can continuously exchange the related information among neighbors via a communication network. Then, inspired by [8], a distributed projected algorithm is proposed to tackle the optimization problem (8.9) as follows Generator $i \in \mathscr{N}_G$:

$$\dot{P}_{i,G} = k_{1i}\big[-2a_i P_{i,G} - b_i + \lambda_i\big]_{P_{i,G}-P_{i,G}^m}^{P_{i,G}^M - P_{i,G}} \tag{8.10a}$$

$$\dot{\lambda}_i = -k_{2i}\sum_{j\in N_i} Y_{ij}(\lambda_i - \lambda_j) - k_{2i}\rho P_{i,G} \tag{8.10b}$$

Load $r \in \mathscr{N}_L$:

$$\dot{P}_{r,L} = k_{1r}\big[-2\alpha_r P_{r,L} + \beta_r - \lambda_r\big]_{P_{r,L}-P_{r,L}^m}^{P_{r,L}^M - P_{r,L}} \tag{8.10c}$$

$$\dot{\lambda}_r = -k_{2r}\sum_{j\in N_r} Y_{rj}(\lambda_r - \lambda_j) + k_{2r}\rho P_{r,L}, \tag{8.10d}$$

where $k_{1i}, k_{2i} > 0$ are control gains, $[v]_{x-c_1}^{c_2-x} = 0$ if $x - c_1 = 0$ and $v \le 0$ or $c_2 - x = 0$ and $v \ge 0$, otherwise $[v]_{x-c_1}^{c_2-x} = v$. In order to deal with the box constraints, a projection operator is introduced in (8.10), where the terms $-2a_i P_{i,G} - b_i + \lambda_i$ and $-2\alpha_r P_{r,L} + \beta_r - \lambda_r$ are projected onto the tangent cones $\mathscr{T}_{\mathscr{F}_i^G}(P_{i,G})$ and $\mathscr{T}_{\mathscr{F}_r^L}(P_{r,L})$, respectively [8]. If the box constraints are relaxed, the projection operator in (8.10) will not be needed.

Remark 8.2 In [4–6, 9], the Lagrangian multiplier λ_i is interpreted as the local selling (or paying) price by generation (or demand) and its global optimal value will eventually converge to the market clearing price p. However, since the transmission costs are included in the optimization problem (8.9), the optimal values of the Lagrangian multipliers λ_i in (8.10 do not converge to a market clearing price p but satisfy (8.14b). As a result, the consensus-based distributed control algorithms in [4–6, 9] will not be applicable for solving the optimization problem (8.9). Therefore, the distributed projected control algorithm (8.10) is developed to deal with the optimization problem (8.9).

Remark 8.3 In real applications, implementing the continuous control algorithm (8.10) may be difficult because continuous monitoring signals are required. On the other hand, it is not possible to continuously change the 'electricity price' λ_i in an electricity market. In practice, one commonly used approach to addressing this issue is that the algorithm (13) is implemented in a discrete way with a small step. In this case, it becomes practical and feasible to update the 'electricity price' λ_i in a fixed time period. It is well acknowledged that the convergence of its discretized algorithm can still be guaranteed when the update step is sufficiently small.

Remark 8.4 In [10], the residual demand derivative can be calculated to construct the best response to competitors' strategies in the presence of transmission constraints. A uniform-price mechanism was designed in [11] to allocate load adjustment to multiple consumers while minimizing the aggregate consumer disutility. It is noted that the market mechanism in [10, 11] primarily focuses on the demand-side management. However, in our design both generation and demand sides are simultaneously considered. In other words, all generators and loads are the market participants who can carry out the energy management and control of smart grids. Moreover, the distributed implementation of our algorithm only needs the local information of the grid, thus improving flexibility and reliability in managing smart grids.

8.2.2 Convergence Analysis

In order to facilitate the convergence analysis, the distributed control algorithm (8.10) is rewritten in a compact form as

$$\dot{P} = K_1\big[-MP - B + \Lambda\big]_{P-P^m}^{P^M-P} \tag{8.11a}$$

$$\dot{\Lambda} = -K_2 L\Lambda - K_2 \rho P \tag{8.11b}$$

where $K_1 = \text{diag}\{k_{1i}\}$, $K_2 = \text{diag}\{k_{2i}\}$, and

$$P^M = [P_{1,G}^M, \ldots, P_{n,G}^M, -P_{n+1,L}^m, \ldots, -P_{n+m,L}^m]^T$$
$$P^m = [P_{1,G}^m, \ldots, P_{n,G}^m, -P_{n+1,L}^M, \ldots, -P_{n+m,L}^M]^T.$$

We are ready to present the following main result.

Theorem 8.1 *Given the initial conditions $P_{i,G}(0) \in \mathscr{F}_i^G$ for $i \in \mathscr{N}_G$ and $P_{r,L}(0) \in \mathscr{F}_r^L$ for $r \in \mathscr{N}_L$, if Assumption 8.1.1 holds, then the distributed control algorithm (8.10) asymptotically converges to the optimal solution to (8.9).*

Proof Define the Lagrangian function

$$\begin{aligned}\mathscr{L}(P,\theta,\Lambda) = &\sum_{i\in\mathscr{N}_G} C_i(P_{i,G}) - \sum_{r\in\mathscr{N}_L} U_r(P_{r,L}) + \frac{\rho}{4}\sum_{i\in\mathscr{N}}\sum_{j\in N_i} Y_{ij}(\theta_i-\theta_j)^2 \\ &+ \sum_{i\in\mathscr{N}_G}\lambda_i\Big(\sum_{j\in N_i} Y_{ij}(\theta_i-\theta_j) - P_{i,G}\Big) \\ &+ \sum_{r\in\mathscr{N}_L}\lambda_r\Big(\sum_{j\in N_r} Y_{rj}(\theta_r-\theta_j) + P_{r,L}\Big)\end{aligned} \tag{8.12}$$

where λ_i is the Lagrangian multiplier and $\Lambda = [\lambda_1, \ldots, \lambda_{n+m}]^T$.

Based on Theorem 3.34 in [1], the Karush-Kuhn-Tucker (KKT) condition for optimization problem (8.9) is derived as

$$0 \in \nabla C_i(P_{i,G}^*) - \lambda_i^* + \mathscr{C}_{\mathscr{F}_i^G}(P_{i,G}^*), \quad i \in \mathscr{N}_G \tag{8.13a}$$

$$0 \in -\nabla U_r(P_{r,L}^*) + \lambda_r^* + \mathscr{C}_{\mathscr{F}_r^L}(P_{r,L}^*), \quad r \in \mathscr{N}_L \tag{8.13b}$$

$$\rho \sum_{j \in N_i} Y_{ij}(\theta_i^* - \theta_j^*) + \sum_{j \in N_i} Y_{ij}(\lambda_i^* - \lambda_j^*) = 0, \quad i \in \mathscr{N} \tag{8.13c}$$

$$P_{i,G}^* = \sum_{j \in N_i} Y_{ij}(\theta_i^* - \theta_j^*), \quad i \in \mathscr{N}_G, \tag{8.13d}$$

$$P_{r,L}^* = -\sum_{j \in N_r} Y_{rj}(\theta_r^* - \theta_j^*), \quad r \in \mathscr{N}_L, \tag{8.13e}$$

$$P_{i,G}^* \in \mathscr{F}_i^G, \quad P_{r,L}^* \in \mathscr{F}_r^L. \tag{8.13f}$$

where $\nabla C_i(P_{i,G}^*) = 2a_i P_{i,G}^* + b_i$ and $\nabla U_r(P_{r,L}^*) = -2\alpha_r P_{r,L}^* + \beta_r$ with $P_{i,G}^*$, $P_{r,L}^*$, θ_i^* and Λ_i^* being the optimal values of $P_{i,G}$, $P_{r,L}$, θ_i and Lagrange multiplier λ_i, respectively. $\mathscr{C}_{\mathscr{F}_i^G}(P_{i,G}^*)$ is the normal cone at the optimal power $P_{i,G}^*$. Then, solving (8.13), one can obtain

$$0 \in MP^* + B - \Lambda^* + \mathscr{C}_{\mathscr{F}}(P^*) \tag{8.14a}$$

$$\rho P^* + L\Lambda^* = 0, \tag{8.14b}$$

$$P^* = L\theta^*, \tag{8.14c}$$

$$P^* \in \mathscr{F}. \tag{8.14d}$$

where $M = \text{diag}\{2a_1, \ldots, 2a_n, 2\alpha_{n+1}, \ldots, 2\alpha_{n+m}\}$, $B = [b_1, \ldots, b_n, \beta_{n+1}, \ldots, \beta_{n+m}]^T$, P^*, θ^* and Λ^* are the optimal value of power P, phasor θ and Lagrange multiplier Λ, respectively. L is the Laplacian matrix of the graph $\mathscr{G}$, which is positive semidefinite. Under Assumption 8.1.1, the kernel of L is of dimension 1. Denote by $(\bar{P}, \bar{\Lambda})$ the equilibrium point of (8.11). Then, by setting $\dot{P} = 0$ and $\dot{\Lambda} = 0$, we obtain that

$$0 = [-M\bar{P} - B + \bar{\Lambda}]_{\bar{P}-P^m}^{P^M-\bar{P}} \tag{8.15a}$$

$$0 = L\bar{\Lambda} + \rho\bar{P} \tag{8.15b}$$

Following similar arguments as in Theorem 4.2 in [8], the projection operator implies that $P_{i,G}(t) \in \mathscr{F}_i^G, i \in \mathscr{N}_G$ and $P_{r,L}(t) \in \mathscr{F}_r^L, r \in \mathscr{N}_L$ for $\forall t \geq 0$ provided that $P_{i,G}(0) \in \mathscr{F}_i^G, i \in \mathscr{N}_G$ and $P_{r,L}(0) \in \mathscr{F}_r^L, r \in \mathscr{N}_L$. It means that the trajectories of (8.10a) and (8.10c) are included in the set $\mathscr{F}_i^G$ and $\mathscr{F}_r^L$, respectively. Since the projected operator makes $-M\bar{P} - B + \bar{\Lambda}$ projected on the cone $\mathscr{C}_{\mathscr{F}}(\bar{P})$, the solution to (8.15) has the same form as (8.14a) and (8.14b). Therefore, it is concluded that the equilibrium point of (8.10) is identical to the optimal solution (8.14).

Next, we analyze the convergence of the distributed control law (8.10) to the optimal values P^* and λ^* in (8.14).

Define the Lyaponov function as

$$V = \frac{1}{2}(P - P^*)^T K_1^{-1}(P - P^*) + \frac{1}{2\rho}(\Lambda - \Lambda^*)^T K_2^{-1}(\Lambda - \Lambda^*). \tag{8.16}$$

According to the characteristic of the projection, it is easy to know that

$$\begin{aligned}&(P - P^*)^T\big[-MP - B + \Lambda\big]_{P-P^m}^{P^M-P}\\ &\le (P - P^*)^T(-MP - B + \Lambda).\end{aligned} \tag{8.17}$$

From (8.14a), one has that $-MP^* - B + \Lambda^* \in \mathscr{C}_{\mathscr{F}}(P^*)$, which implies that, for all $P \in \mathscr{F}$,

$$(P - P^*)^T(-MP^* - B + \Lambda^*) \le 0. \tag{8.18}$$

Combining (8.17) with (8.18), one obtains that

$$\begin{aligned}&(P - P^*)^T\big[-MP - B + \Lambda\big]_{P-P^m}^{P^M-P}\\ &\le (P - P^*)^T(-MP + MP^* + \Lambda - \Lambda^*).\end{aligned} \tag{8.19}$$

Taking the derivative of V along (8.10) and using (8.19) yield

$$\begin{aligned}\dot{V} &= (P - P^*)^T\big[-MP - B + \Lambda\big]_{P-P^m}^{P^M-P}\\ &\quad + \frac{1}{\rho}(\Lambda - \Lambda^*)^T(-L\Lambda - \rho P)\\ &\le -(P - P^*)^T M(P - P^*)\\ &\quad - \frac{1}{\rho}(\Lambda - \Lambda^*)^T L(\Lambda - \Lambda^*).\end{aligned} \tag{8.20}$$

Since $M > 0$ and the Laplacian matrix L is positive semidefinite, we have $\dot{V} \le 0$. To apply the invariant set theorem, we define $\mathscr{S} = \{(P, \Lambda) | \dot{V} = 0\}$. Since $M > 0$ and under Assumption 8.1.1 the kernel of L is one dimensional, $\mathscr{S} = \{(P, \Lambda) | P = P^*, \Lambda - \Lambda^* = \nu v, \nu \in R\}$, where v is a non-zero constant vector in the kernel of L. Furthermore, for any $(P, \Lambda) \in \mathscr{S}$ which is not the equilibrium point of (8.10), by (8.11), $\dot{P} \neq 0$, implying that the state will leave $\mathscr{S}$. Consequently, the largest invariant set in $\mathscr{S}$ is the equilibrium point of (8.10). Thus, based on the LaSalle's invariant set theorem in Lemma 8.1, it is concluded that the trajectories (8.10) will converge to the optimal solution P^* and Λ^* in (8.14). The proof is complete. □

Remark 8.5 In the literature [12, 13], it is required that the total balance of the initial generation power $P_{i,G}(0)$ and demand power $P_{r,L}(0)$ is satisfied, i.e.,

$\sum_{i \in \mathcal{N}_G} P_{i,G}(0) = \sum_{r \in \mathcal{N}_L} P_{r,L}(0)$. In reality, such requirement is very restrictive, especially, in intermittent generation networks that involve frequent connection and disconnection of generators. By contrast, it is more practical to implement the distributed control algorithm (8.10) which only requires that $P_{i,G}(0) \in \mathcal{F}_i^G$ and $P_{r,L}(0) \in \mathcal{F}_r^L$.

Remark 8.6 Recently, various distributed optimization algorithms have been designed to address the energy management issue. Reference [14] proposed a decoupling algorithm to perform the demand side management, where the central utility company is required to collect total demands from all customers over a communication network. In [15], a fully distributed optimization algorithm for demand response was presented based on Predictor Corrector Proximal Multiplier (PCPM). Compared with the distributed algorithm in [14, 15], the advantages of our distributed algorithms (8.10) include that: (i) it requires only local communication between neighbor buses rather than one-to-all communication of [14]; (ii) the distributed algorithm in [15] needs to solve the optimization subproblem at each iteration, which may result in an overload of computation; However, such complicated computation can be avoided by our proposed algorithm (8.10); (iii) it is easy to conduct the convergence analysis of the algorithm (8.10).

8.3 Distributed Event-Triggered Control Algorithms

The distributed control algorithm in the previous section assumes instantaneous information exchange. In practice, such distributed control algorithm (8.10) must be implemented by a sampled-data communication scheme. The traditional uniform sampling scheme uses a fixed sampling period T. When T is small, communication data flow rates become high, causing potential network congestion. Lots of communication resources may be wasted if signal variations between two consecutive sampling instances are small. With limited communication resources such as network bandwidth and communication channels, it is essential to design a suitable communication scheme which can significantly reduce communication resource consumption while sustaining satisfactory control performance. It has been verified by [16–18] that the event-triggered scheme has outstanding advantages over traditional periodic scheme in saving the utilization of communication resources. Therefore, we employ an event-triggered communication scheme to reduce communication data flow in this section.

Note that, in the distributed optimization control algorithm (8.10) the measurement of Lagrangian multipliers λ_i is required to be exchanged among neighbors via a communication network. Thus, we only need to consider the sampling and communication action of λ_i. For the clarity of presentation, the framework of our proposed event-triggered control scheme is shown in Fig. 8.1. To implement the event-triggered communication scheme, an event detector i, in which an event-triggered condition (ETC) is preset and embedded, is introduced and installed for unit i. It is assumed that the generation (or demand) power $P_{i,G}$ (or $P_{i,L}$) and the variable λ_i can be monitored

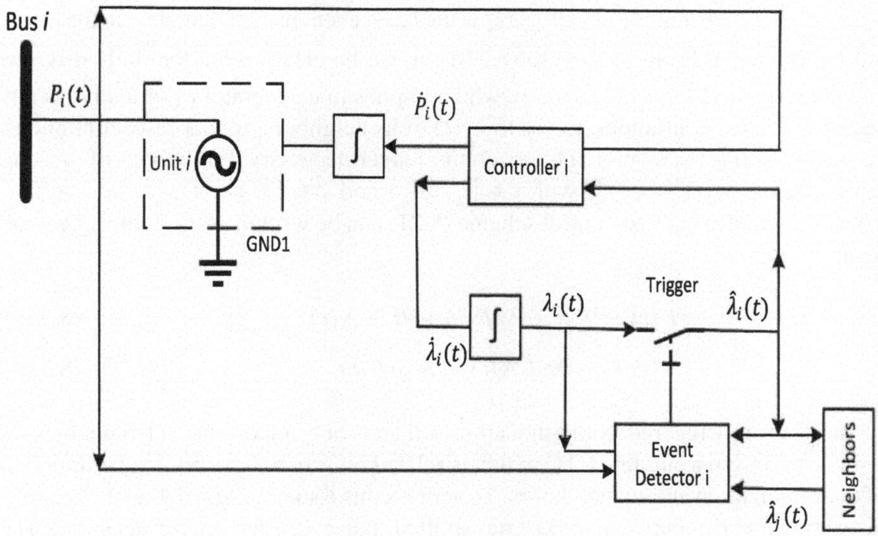

Fig. 8.1 A framework of the event-triggered control scheme at unit i

continuously and sent to event detector i. The fundamental task of event detector i is to determine when the measurement of λ_i should be transmitted to its neighbors based on its predefined ETC.

To be specific, let the sequence $T_i \triangleq \{t_k^i \mid k = 0, 1, \ldots, \infty\}, i \in \mathcal{N}$ denote the event-triggered instants with $0 = t_0 < t_1^i < t_2^i \cdots < t_k^i < \cdots$. If the ETC of generation i (or load r) is satisfied at each instant t_k^i, the variable λ_i of node i will be sampled, and then the sampled-data $\lambda_i(t_k^i)$ of node i will be broadcast to its neighbors. It means that only the measurements at the event-triggered sampling instants are available to the controller update of itself and its neighbors. A zero order hold (ZOH) is used to keep the measurement unchanged for the controller update in the interval $t \in [t_k^i, t_{k+1}^i)$ until the next event-triggered state is received. Based on this scheme, we propose a distributed event-triggered control algorithm for $t \in [t_k^i, t_{k+1}^i)$ as follows.

Generator $i \in \mathcal{N}_G$:

$$\dot{P}_{i,G}(t) = k_{1i}\big[-2a_i P_{i,G}(t) - b_i + \lambda_i(t_k^i)\big]_{P_{i,G}(t)-P_{i,G}^m}^{P_{i,G}^M - P_{i,G}(t)}, \tag{8.21a}$$

$$\dot{\lambda}_i(t) = -k_{2i} \sum_{j \in N_i} Y_{ij}(\lambda_i(t_k^i) - \lambda_j(t_{k_j(t)}^j)) - k_{2i}\rho P_{i,G}(t), \tag{8.21b}$$

Load $r \in \mathcal{N}_L$:

$$\dot{P}_{r,L}(t) = k_{1r}\big[-2\alpha_r P_{r,L}(t) + \beta_r - \lambda_r(t_k^r)\big]_{P_{r,L}(t)-P_{r,L}^m}^{P_{r,L}^M - P_{r,L}(t)}, \tag{8.21c}$$

$$\dot{\lambda}_r(t) = -k_{2r} \sum_{j \in N_r} Y_{rj}(\lambda_r(t_k^r) - \lambda_j(t_{k_j(t)}^j)) + k_{2r}\rho P_{r,L}(t), \tag{8.21d}$$

where $t^j_{k_j(t)} \triangleq \arg\min_{t^j_p}\{t - t^j_p | t \geq t^j_p\}$ is the latest event instant of node j in the interval $t \in [t^i_k, t^i_{k+1})$. From (8.21b) to (8.21d), it can be clearly seen that only discrete signals $\lambda_i(t^i_k)$ and $\lambda_j(t^j_{k_j(t)})$ are used, which implies that generator i (or load r) is not required to have continuous access to $\lambda_j(t)$ of its neighbors. In this case, continuous communications between neighbors are no longer necessary. For brevity of presentation, define $\hat{\lambda}_i(t) \triangleq \lambda_i(t^i_k)$ with $t \in [t^i_k, t^i_{k+1})$ and $\hat{\Lambda}(t) \triangleq [\hat{\lambda}_1(t), \ldots, \hat{\lambda}_{n+m}(t)]^T$. Then, the event-triggered control scheme (8.21) can be written as the compact vector form

$$\dot{P}(t) = K_1\big[-MP(t) - B + \hat{\Lambda}(t)\big]^{P^M - P(t)}_{P(t)-P^m}, \tag{8.22a}$$

$$\dot{\Lambda}(t) = -K_2 L\hat{\Lambda}(t) - K_2\rho P(t). \tag{8.22b}$$

In an event-triggered communication scheme, one important and fundamental issue is to design a suitable ETC which is related closely to determining the number of communication among neighbors. To address this issue, we first define the relative measurement error between the last transmitted state $\lambda_i(t^i_k)$ and the current state $\lambda_i(t)$ of node i as

$$e_i(t) \triangleq \lambda_i(t^i_k) - \lambda_i(t), \quad t \in [t^i_k, t^i_{k+1}). \tag{8.23}$$

Then, define the triggering functions as

$$f_{i,G}(t) = |e_i| - \phi_i(t), \qquad f_{r,L}(t) = |e_r| - \phi_r(t),$$

where

$$\phi_i(t) = \eta_i\left|\sum_{j\in N_i} Y_{ij}(\hat{\lambda}_i(t) - \hat{\lambda}_j(t)) + \rho P_{i,G}(t)\right|,$$

$$\phi_r(t) = \eta_r\left|\sum_{j\in N_i} Y_{rj}(\hat{\lambda}_r(t) - \hat{\lambda}_j(t)) - \rho P_{r,L}(t)\right|$$

with $\eta_i > 0, i \in \mathcal{N}$ being the threshold parameters. Here, $\phi_i(t)$ is called a threshold function of error $e_i(t)$. Hence, the event instant t^i_{k+1} is determined by the following rule

$$t^i_{k+1} = \begin{cases} \inf\limits_{t>t^i_k, i\in\mathcal{N}_G} \{t \mid f_{i,G}(t) > 0\} \\ \inf\limits_{t>t^r_k, r\in\mathcal{N}_L} \{t \mid f_{r,L}(t) > 0\} \end{cases} \tag{8.24}$$

Clearly, (8.24) establishes the event-triggered communication scheme. It is noted that the ETC in (8.24) is checked in a distributed manner since only the local measurement information of error $e_i(t)$ and threshold $\phi_i(t)$ are used. Based on the scheme

(8.24), it is known that, once the ETC in (8.24) is satisfied at event instant t_k^i, the related variable information λ_i of node i at t_k^i will be sampled and then be allowed to be broadcast to its neighbors. A reasonable explanation to the ETC in (8.24) is that the change $e_i(t)$ of λ_i between the consecutive event instants is not allowed to exceed the threshold $\phi_i(t)$. In other words, the new sampling and communication action of λ_i only occurs when its relative measurement errors exceed the allowable range. In this sense, the number of interactive communications among neighboring units can be significantly reduced. It should be pointed out that $e_i(t)$ will be reset as zero after the ETC is triggered at the event times t_k^i.

In what follows, we establish the following conclusion.

Theorem 8.2 *Given the initial conditions $P_{i,G}(0) \in \mathscr{F}_i^G$ for $i \in \mathscr{N}_G$ and $P_{r,L}(0) \in \mathscr{F}_r^L$ for $r \in \mathscr{N}_L$, if Assumption 8.1.1 holds and the threshold parameters satisfy*

$$0 < \eta_i < \min\left\{\frac{\sigma_{\min}(M)}{2\rho}, \frac{1}{2\sigma_{\max}(L)}\right\},$$

then the distributed control algorithm (8.10) with the event-triggered communication scheme (8.24) asymptotically converges to the optimal solution to (8.9). Moreover, no Zeno behavior happens under the event-triggered communication scheme (8.24).

Proof Choose the same candidate of the Lypunov function as (8.16). Taking derivatives of $V(t)$ along with (8.22) yields

$$\begin{aligned}
\dot{V}(t) =& (P(t) - P^*)^T\big[-MP(t) - B + \hat{\Lambda}(t)\big]_{P(t)-P^m}^{P^M-P(t)} \\
&+ \frac{1}{\rho}(\Lambda(t) - \Lambda^*)^T(-L\hat{\Lambda}(t) - \rho P(t)) \\
\leq& (P(t) - P^*)^T(-MP(t) + MP^* + \hat{\Lambda}(t) - \Lambda^*) \\
&+ \frac{1}{\rho}(\hat{\Lambda}(t) - \Lambda^* + e(t))(-L\hat{\Lambda}(t) - \rho P(t)) \\
\leq& -(P(t) - P^*)^T M(P(t) - P^*) - \frac{1}{\rho}(\hat{\Lambda}(t) - \Lambda^*)L(\hat{\Lambda}(t) - \Lambda^*) \\
&+ \frac{1}{\rho}\|e(t)\|\|L\hat{\Lambda}(t) + \rho P(t))\|
\end{aligned} \tag{8.25}$$

where $e(t) \triangleq [e_1(t), \ldots, e_{n+m}(t)]^T$ and $\lambda(t) = \hat{\lambda}(t) - e(t)$ is used in the second inequality.

Based on the event-triggered communication scheme (8.24), the following inequalities $f_{i,G}(t) \leq 0$ and $f_{r,L}(t) \leq 0$ hold for $t \in [t_k^i, t_{k+1}^i)$, which implies that

$$\|e(t)\| \leq \eta_{\max}\|L\hat{\Lambda}(t) + \rho P(t)\| \tag{8.26}$$

where $\eta_{\max} \triangleq \max_{i \in \mathscr{N}}\{\eta_i\}$. Then, one can get

$$
\begin{aligned}
&\|e(t)\|\|L\hat{\Lambda}(t)+\rho P(t))\| \\
&\le \eta_{\max}\|L\hat{\Lambda}(t)+\rho P(t)\|^2 \\
&\le 2\eta_{\max}\rho^2(P(t)-P^*)^T(P(t)-P^*) \\
&\quad + 2\eta_{\max}(\hat{\Lambda}(t)-\Lambda^*)LL(\hat{\Lambda}(t)-\Lambda^*) \\
&\le 2\eta_{\max}\rho^2(P(t)-P^*)^T(P(t)-P^*) \\
&\quad + 2\eta_{\max}\sigma_{\max}(L)(\hat{\Lambda}(t)-\Lambda^*)L(\hat{\Lambda}(t)-\Lambda^*). \qquad (8.27)
\end{aligned}
$$

Substituting (8.27) into (8.25) results in

$$
\begin{aligned}
\dot{V}(t) \le & -(P(t)-P^*)^T(M-2\eta_{\max}\rho I_{n+m})(P(t)-P^*) \\
& -\frac{1}{\rho}(1-2\eta_{\max}\sigma_{\max}(L))(\hat{\Lambda}(t)-\Lambda^*)L(\hat{\Lambda}(t)-\Lambda^*).
\end{aligned}
$$

Since $0<\eta_i<\min\{\frac{\sigma_{\min}(M)}{2\rho}, \frac{1}{2\sigma_{\max}(L)}\}$, it is guaranteed that $\dot{V}(t)\le 0$. Similar to the proof of Theorem 8.1, using LaSalle's invariance principle in Lemma 8.1, it is concluded that the event-triggered control scheme (8.21) will converge to the optimal solution P^* and Λ^* in (8.14).

Next, we will analyze that the event-triggered communication scheme does not exhibit the Zeno behavior, namely, no infinite communication happens to each unit in any finite time period. In order to address this issue, we will show that the lower bound of inter-event times are strictly larger than zero.

Assume that unit i triggers at time t_k^i, then we have $e_i(t_k^i)=0$. For $t\in[t_k^i, t_{k+1}^i)$, it is clear that $\dot{e}_i(t)=-\dot{\lambda}_i(t)$. If $\phi_i(t)=0$, we have $\dot{e}(t)=0$ and $e(t)=0$ for $t\ge t_k^i$. In this case, no broadcasts will occur based on (8.24). Hence, we only consider the case of $\phi_i(t)\ne 0$, namely, $\phi_i(t)>0$. Based on (8.21a) and the event condition (8.24), one can have for $t\in[t_k^i, t_{k+1}^i)$

$$
\begin{aligned}
\eta_i\phi_i(t) < |e_i(t)| &\le k_{2,i}\int_{t_k}^{t}\phi_i(s)ds \\
&\le k_{2,i}(t-t_k^i)(|\hat{z}_i|^{\max}+\rho|P_i|^{\max}) \qquad (8.28)
\end{aligned}
$$

where $|\hat{z}_i|^{\max}$ represents the maximum value of $\hat{z}_i(t)$ for $t\in[t_k^i, t_{k+1}^i)$ with

$$
\hat{z}_i(t)=\sum_{j\in N_i} Y_{ij}(\hat{\lambda}_i(t)-\hat{\lambda}_j(t)),
$$

and $|P_i|^{\max}=P_{i,G}^M$ for $i\in\mathcal{N}_G$ or $|P_i|^{\max}=P_{i,L}^M$ for $i\in\mathcal{N}_L$. Thus, the next event time t_{k+1}^i will satisfy the following inequality

$$
\eta_i\phi_i^{\min} < k_{2,i}(t_{k+1}^i-t_k^i)(|\hat{z}_i|^{\max}+\rho|P_i|^{\max}) \qquad (8.29)
$$

where $\phi_i^{\min} > 0$ is the minimum value of $\phi_i(t)$ for $t \in [t_k^i, t_{k+1}^i]$. Then, we have

$$t_{k+1}^i - t_k^i > \frac{\eta_i \phi_i^{\min}}{k_{2,i}(|\hat{z}_i|^{\max} + \rho |P_i|^{\max})} > 0. \tag{8.30}$$

It is clear from (8.30) that the lower bound of the inter-event times is strictly positive, which implies that the Zeno behavior is excluded. The proof of Theorem 8.2 is complete. □

Remark 8.7 The penalty term $\frac{\rho}{4}\sum_{i\in\mathcal{N}}\sum_{j\in N_i} P_{ij}^2/Y_{ij}$ in our optimization formulation (8.9) is used to limit the transmission line powers. Then, we establish the optimality of distributed strategies and convergence of our distributed algorithms under event-triggered communication. Technically, it is certainly possible to impose capacity constraints on the transmission line powers. In principle, optimization algorithms will then involve further projection operators from these bounds (currently, the projection operator on the bus powers is only utilized). It will be a very meaningful study to explore such alternative formulations, which will be further discussed and carefully addressed in our future work.

Remark 8.8 (*Computation of optimal phase angle* θ^*) Note that since the matrix L is the Laplacian matrix with its kernel being of dimension 1 under Assumption 8.1.1, the optimal solution θ^* to the DC power flow $P^* = L\theta^*$ is not unique when the optimal power injection P^* is known. In order to address this issue, without loss of generality, we set $\theta_1^* = 0$ as a reference phase angle. By solving (8.14b) and (8.14c), namely $L(\rho\theta^* + \Lambda^*) = 0$, one gets $\theta^* = -\frac{1}{\rho}\Lambda^* + \nu\mathbf{1}$, where ν is a constant. Then it is easy to obtain that $\nu = \frac{1}{\rho}\lambda_1^*$ due to $\theta_1^* = 0$. As a consequence, the optimal phase angle can be calculated by $\theta^* = -\frac{1}{\rho}\Lambda^* + \frac{1}{\rho}\lambda_1^*\mathbf{1}$. It should be pointed out that each unit can have access to the value of ν among the neighboring units via a communication network.

8.4 Simulation Case Studies

In this section, some simulation tests are conducted to show the effectiveness of the distributed control algorithm proposed in this chapter. We consider the IEEE 9-bus system with three generators and six loads, which is shown in Fig. 8.2.

In this case study, the related parameters of transmission lines are listed in Table 8.1. From Table 8.1, it is clearly observed that the resistance of each transmission line is much smaller than its reactance,[1] thus it is reasonable to assume that the resistance is ignored, which implies that the active power loss is negligible and will not be included in control strategy consideration. In this case, the DC power

[1] In our simulation using Matlab/PSAT, the small power loss is compensated by introducing a slack bus.

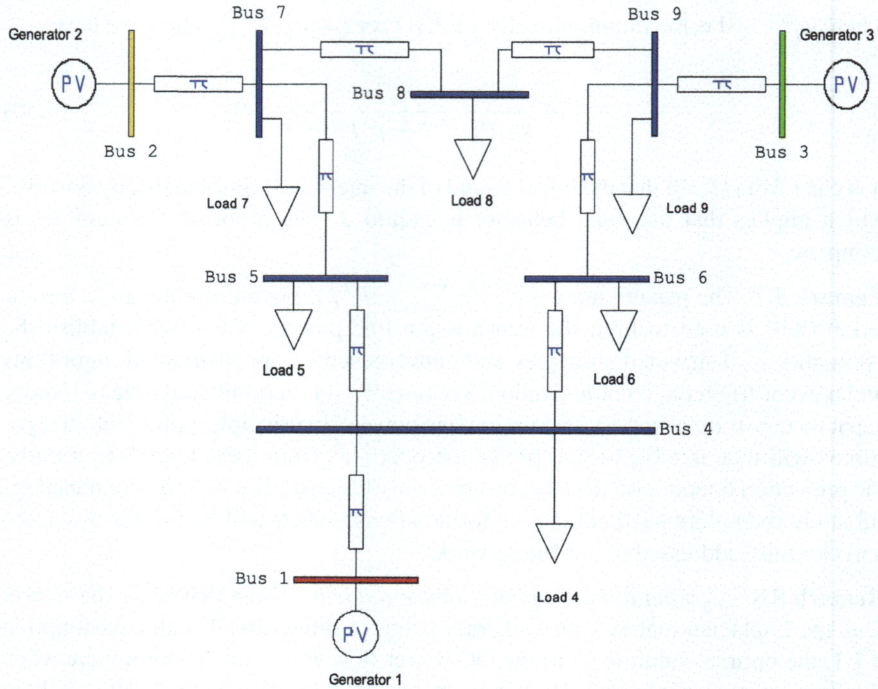

Fig. 8.2 IEEE 9-bus system with three generators and six loads

Table 8.1 Cost coefficients for generators and loads

Transmission line	Reactance	Resistance
Bus 1 ⟷ Bus 4	0.850	0.039
Bus 2 ⟷ Bus 7	0.360	0.0085
Bus 3 ⟷ Bus 9	0.360	0.0085
Bus 4 ⟷ Bus 5	0.425	0.010
Bus 4 ⟷ Bus 6	0.460	0.0170
Bus 5 ⟷ Bus 7	0.805	0.032
Bus 6 ⟷ Bus 9	0.850	0.039
Bus 7 ⟷ Bus 8	0.360	0.0085
Bus 8 ⟷ Bus 9	0.504	0.0119

flow model can be used. Correspondingly, we can compute the matrix L easily. The related parameters of the generators and loads are given in Table 8.2. The initializations of the generators and loads are also given in Table 8.2. It is clear that the total supply-demand power is not balanced and the initialization mismatch between the generators and loads is -35 MW. Next, we will demonstrate the efficiency of the proposed distributed control algorithms with both continuous and event-triggered communications, respectively.

Table 8.2 Cost coefficients for generators and loads

Generator	a_i	b_i	$P_{i,G}^M$ (MW)	$P_{i,G}^m$ (MW)	$P_{i,G}(0)$ (MW)
1	0.08	2.25	70	10	40
2	0.062	4.2	65	20	25
3	0.075	3.25	70	30	35
Load	α_r	β_r	$P_{r,L}^M$ (MW)	$P_{r,L}^m$ (MW)	$P_{r,L}(0)$ (MW)
4	0.072	8.25	50	10	10
5	0.066	7.90	60	2	30
6	0.070	7.55	30	3	15
7	0.055	8.00	40	5	25
8	0.075	7.75	40	15	20
9	0.045	8.05	70	5	35

8.4.1 Case 1: Continuous Communication Schemes

Here, it is assumed that the communication network is ideal, where the information transmission can be regarded as being continuous and instantaneous. The simulation time step is $T = 0.01$s. Based on the distributed control algorithm (8.10) with $\rho = 0.2$, $k_{1i} = 40$ and $k_{2i} = 1$ for $i = 1, \ldots, 9$, the evolution of all the generators and loads is plotted in Fig. 8.3 and the total power mismatch is plotted in Fig. 8.4. From Fig. 8.3, it is clearly seen that the output powers of all the generators and loads can eventually converge to the optimal values. Notice that the initial powers of the generators and loads are unbalanced. As shown in Fig. 8.4, using the control algorithm (8.10), the power error between the generators and loads will converge to

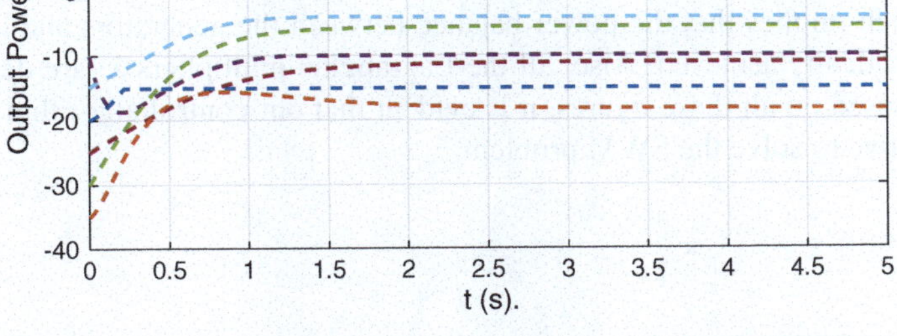

Fig. 8.3 Local generator and load powers with continuous communication network

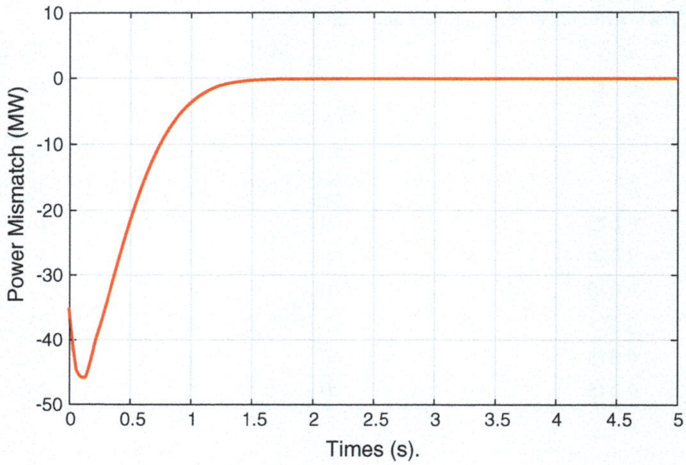

Fig. 8.4 Mismatch between total supply and demand powers with continuous communication network

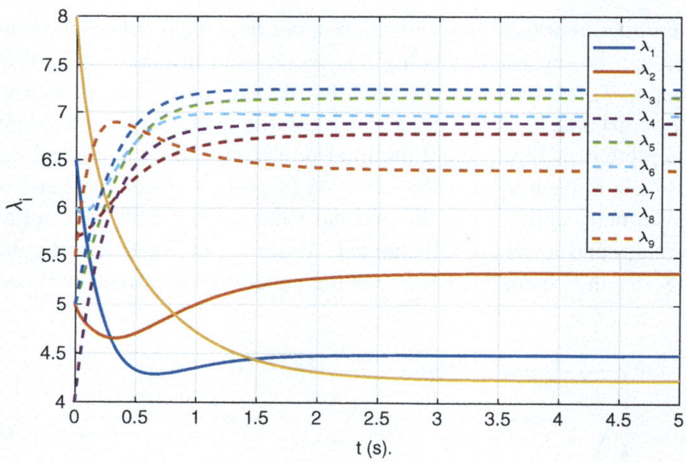

Fig. 8.5 Lagrangian multipliers with continuous communication network

zero,[2] which implies that the power balance between the generators and loads can be asymptotically achieved. Also, all the Lagrangian multipliers λ_i are depicted in Fig. 8.5. Based on all these figures, it is evident that our control algorithm (8.10) is very effective to solve the SWM problem.

[2] In fact, there is about 0.0218 MW active power loss compensation from the slack bus.

8.4.2 Case 2: Event-Triggered Communication Schemes

We now validate the distributed event-triggered control algorithm (8.21) and show its advantages of reducing the frequency of communication over the periodical one (8.10). First, select the same control parameters k_{1i}, k_{2i} and simulation step T as that of Case 1. For simplicity, set the threshold parameter $\eta_i = 0.0491$ for all $i = 1, \ldots, 9$. From the ETC (8.24), it is known that when output power of each unit approaches the equilibrium state, both $e_i(t)$ and $\phi_i(t)$ will become very small. It results in that more events may be triggered, especially, in the presence of measurement disturbances. On the other hand, in many applications it may be only required that the control accuracy is achieved within an allowable degree. Based on this consideration, the threshold of ETC (8.24) is modified as $\max\{\phi_i(t), \epsilon\}$, where the parameter ϵ represents the control accuracy. In this case, the convergence of the algorithm can be still guaranteed. Here, set $\epsilon = 10^{-5}$ for the case study.

Figures 8.6, 8.7 and 8.8 demonstrate the trajectories of the output powers of the generators and loads, power mismatch, and increment costs. Compared with those shown in Figs. 8.3, 8.4 and 8.5, it is evident that there is little performance loss under the event-triggered communication. The event-triggered instants and release intervals of data transmission at generator 1 and load 8 are shown in Fig. 8.9. From the y-axis of Fig. 8.9, it is obvious that the communication intervals are aperiodic, for example, the maximum communication interval at load 8 is approximately 0.2 s. It also implies that communication exchange happens when "requested".

To show the advantages of the event-triggered communication scheme over the periodic one, we make a comparison by using the event-triggered control algorithm (8.21) and that of Case 1. The numbers of communication under these two control

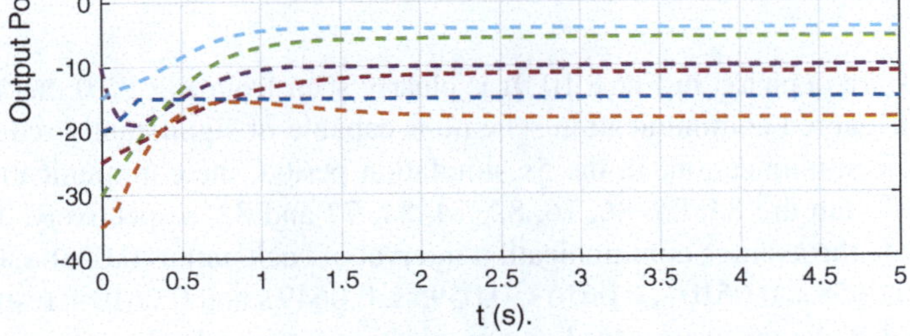

Fig. 8.6 Local generator and load powers with event-triggered communication network

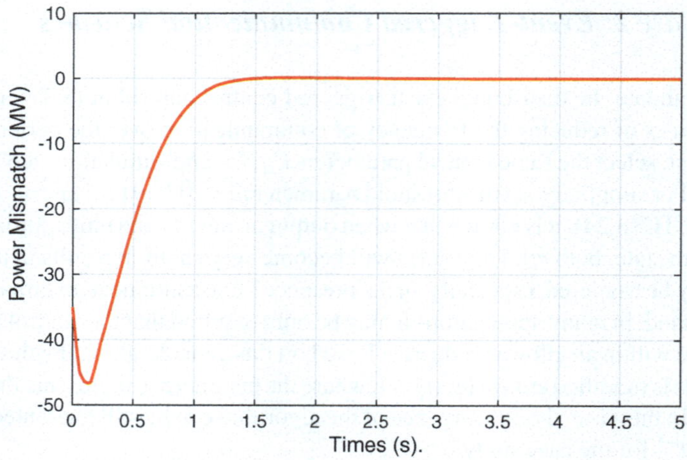

Fig. 8.7 Mismatch between total supply and demand powers with event-triggered communication network

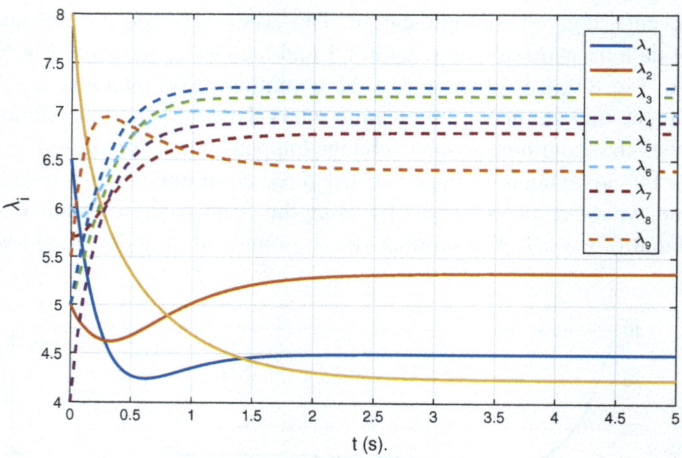

Fig. 8.8 Lagrangian multipliers with event-triggered communication network

algorithms are depicted in Fig. 8.10. It is clearly seen from Fig. 8.10 that adopting the event-triggered communication scheme is capable of significantly reducing the number of communication. In the 5s simulation period, the communication numbers of each unit are 85, 85, 86, 76, 82, 74, 84, 77 and 82, respectively. Then, we can calculate the average communication interval of each unit as 0.0588 s, 0.0588 s, 0.0581 s, 0.0658 s, 0.0610 s, 0.0676 s, 0.0595 s, 0.0649 s and 0.0610 s, respectively. It implies that the event-triggered communication scheme leads to larger average communication intervals than that of Case 1. Moreover, from the viewpoint of communication resource saving, we can calculate that only about 16.09% (1448/9000)

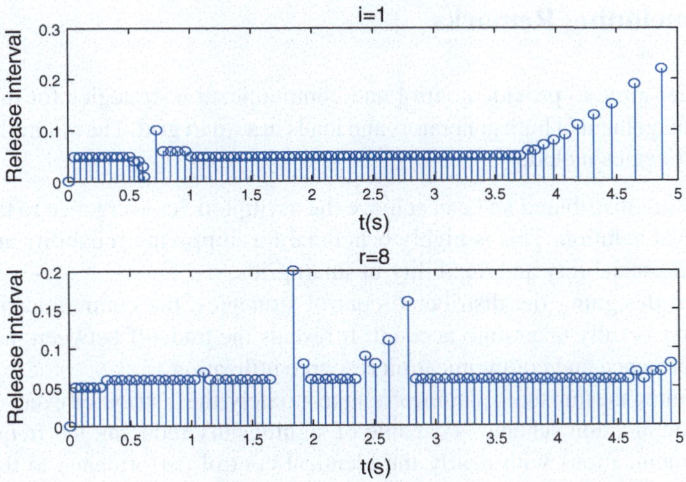

Fig. 8.9 Event instants and release intervals for generator 1 and load 8 under ETC (8.24) with threshold parameters $\eta_1 = \eta_8 = 0.0491$

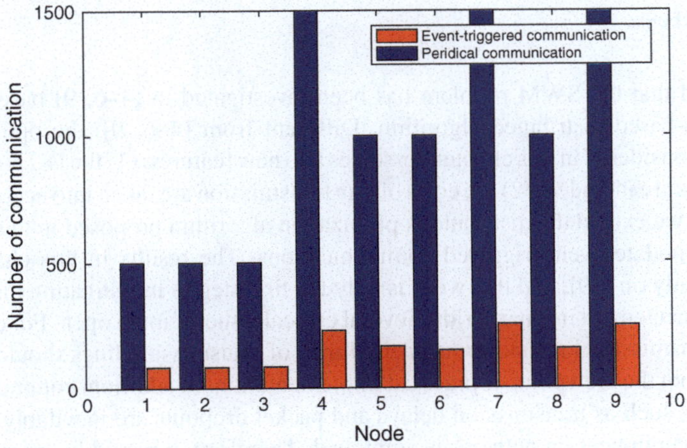

Fig. 8.10 Comparisons of communication number between the event-triggered and periodic communication schemes for each unit

communication resource of Case 1 is needed in the case of event-triggered communications. Based on the above observations, it can be concluded that the distributed control algorithm (8.21) with event-triggered communication scheme (8.24) can achieve nearly the same control performance as the periodic counterpart by only utilizing less communication resource.

8.5 Concluding Remarks

This chapter aims to provide control and communication strategies for distributed energy management of both generators and loads in a smart grid. The main advantages of these strategies include:

(i) They are distributed and can achieve the asymptotic convergence to the global optimal solution. This is highly beneficial for improving reliability and maintaining scalability and flexibility in smart grids.
(ii) When designing the distributed control strategies, the communication complexity is fully taken into account. It reveals the tradeoff between the control performance and communication resource utilization.
(iii) By tuning the threshold parameters appropriately, the distributed event-triggered communication scheme is capable of significantly reducing the frequency of communications with nearly the identical control performance to that under continuous data communications.

8.6 Notes

It is noted that the SWM problem has been investigated in [4–6, 9] based on the consensus-based distributed algorithm. Different from [4–6, 9], the optimization problem considered in this chapter possesses two new features: (1) the DC power flow constraints are included; (2) the costs of line transmission are taken into account. Furthermore, we extend the distributed optimization algorithm proposed in this chapter to accommodate event-triggered communications. The results in this chapter are based mainly on [19], and this work is only the first step in investigating distributed energy management in smart grids. Several critical issues remain open. For example, power transmission loss and the capacity limits of transmission lines should be considered from the viewpoint of practical implementation. In addition, communication constraints such as transmission delays and packet dropouts are inevitably induced when a communication network is employed. Therefore, it would be interesting to explore comprehensive impacts of such communication constraints on the stability and performance of smart grids.

References

1. A.P. Ruszczyński, *Nonlinear Optimization* (Princeton University Press, Princeton, NJ, 2006)
2. H.K. Khalil, *Nonlinear Systems*, 2nd edn. (Prentice-Hall, Upper Saddle River, NJ, 1996)
3. F. Wu, P. Varaiya, P. Spiller, S. Oren, Folk theorems on transmission access: proofs and counterexamples. J. Regul. Econ. **10**(1), 5–23 (1996)

4. N. Rahbari-Asr, U. Ojha, Z. Zhang, M.-Y. Chow, Incremental welfare consensus algorithm for cooperative distributed generation/demand response in smart grid. IEEE Trans. Smart Grid **5**(6), 2836–2845 (2014)
5. W. Zhang, Y. Xu, W. Liu, C. Zang, H. Yu, Distributed online optimal energy management for smart grids. IEEE Trans. Ind. Inform. **11**(3), 717–727 (2015)
6. Y. Xu, Z. Li, Distributed optimal resource management based on the consensus algorithm in a microgrid. IEEE Trans. Ind. Inform. **62**(4), 2584–2592 (2015)
7. P. Samadi, H. Mohsenian-Rad, R. Schober, V.W. Wong, Advanced demand side management for the future smart grid using mechanism design. IEEE Trans. Smart Grid **3**(3), 1170–1180 (2012)
8. P. Yi, Y. Hong, F. Liu, Initialization-free distributed algorithms for optimal resource allocation with feasibility constraints and application to economic dispatch of power systems. Automatica **74**, 259–269 (2016)
9. C. Zhao, J. He, P. Cheng, J. Chen, Consensus-based energy management in smart grid with transmission losses and directed communication. IEEE Trans. Smart Grid **8**(5), 2049–2061 (2017)
10. L. Xu, R. Baldick, Transmission-constrained residual demand derivative in electricity markets. IEEE Trans. Power Syst. **22**(4), 1563–1573 (2007)
11. Y. Xu, N. Li, S.H. Low, Demand response with capacity constrained supply function bidding. IEEE Trans. Power Syst. **31**(2), 1377–1394 (2016)
12. A. Cherukuri, J. Cortés, Distributed generator coordination for initialization and anytime optimization in economic dispatch. IEEE Trans. Control Netw. Syst. **2**(3), 226–237 (2015)
13. W. Yu, C. Li, X. Yu, G. Wen, J. Lü, Distributed consensus strategy for economic power dispatch in a smart grid, in *Proceedings of the 2015 10th Asian Control Conference (ASCC)*, Kota Kinabalu, Malaysia (2015), pp. 1–6
14. N. Li, L. Chen, S. H. Low, Optimal demand response based on utility maximization in power networks, in *Proceedings of 2011 IEEE Power and Energy Society General Meeting*, Detroit, MI (2011), pp. 1–8
15. N. Li, L. Chen, S. H. Low, Demand response in radial distribution networks: Distributed algorithm, in *Proceedings of the 46th Asilomar Conference on Signal Systems Computer*, Pacific Grove, CA (2012), pp. 1549–1553
16. D. Yue, E. Tian, Q.-L. Han, A delay system method for designing event-triggered controllers of networked control systems. IEEE Trans. Autom. Control **58**(2), 475–481 (2013)
17. D. Dimarogonas, E. Frazzoli, K. Johansson, Distributed event-triggered control for multi-agent systems. IEEE Trans. Autom. Control **57**(5), 1291–1297 (2012)
18. G. Guo, L. Ding, Q.-L. Han, A distributed event-triggered transmission strategy for sampled-data consensus of multi-agent systems. Automatica **50**(5), 1489–1496 (2014)
19. L. Ding, Q.-L. Han, L. Wang, E. Sindi, Distributed cooperative optimal control of DC microgrids with communication delays. IEEE Trans. Ind. Inform. **14**(9), 3924–3935 (2018)

Index

L. Ding et al., *Distributed Control and Optimization of Networked Microgrids*, Power Systems, https://doi.org/10.1007/978-3-030-95029-3

The manufacturer's authorised representative in the EU is Springer Nature Customer Service Centre GmbH, Europaplatz 3, 69115 Heidelberg, Germany. If you have any concerns regarding our products, please contact ProductSafety@springernature.com

Printed and bound by CPI Group (UK) Ltd, Croydon, CR0 4YY
22/09/2026
02221055-0020